Lecture Notes in Computer Science 708

Edited by G. Goos and J. Hartmanis

Christian Laugier (Ed.)

Geometric Reasoning for Perception and Action

Workshop
Grenoble, France, September 16-17, 1991
Selected Papers

Springer-Verlag
Berlin Heidelberg New York
London Paris Tokyo
Hong Kong Barcelona
Budapest

Christian Laugier (Ed.)

Geometric Reasoning for Perception and Action

Workshop
Grenoble, France, September 16-17, 1991
Selected Papers

Springer-Verlag
Berlin Heidelberg New York
London Paris Tokyo
Hong Kong Barcelona
Budapest

Series Editors

Gerhard Goos
Universität Karlsruhe
Postfach 69 80
Vincenz-Priessnitz-Straße 1
D-76131 Karlsruhe, Germany

Juris Hartmanis
Cornell University
Department of Computer Science
4130 Upson Hall
Ithaca, NY 14853, USA

Volume Editor

Christian Laugier
LIFIA/INRIA Rhône-Alpes
46, avenue Félix Viallet, F-38031 Grenoble Cédex 1, France

CR Subject Classification (1991): I.3.5, I.3.7, I.2.9-10

ISBN 3-540-57132-9 Springer-Verlag Berlin Heidelberg New York
ISBN 0-387-57132-9 Springer-Verlag New York Berlin Heidelberg

© Springer-Verlag Berlin Heidelberg 1993
Printed in Germany

Typesetting: Camera-ready by authors
45/3140-543210 - Printed on acid-free paper

Preface

Geometry is a powerful tool to solve a great number of problems in robotics and computer vision. Impressive results have been obtained in these fields in the last decade. However addressing problems of the actual world requires reasoning about *uncertainty* and *complex motion constraints* by combining geometric, kinematic, and dynamic characteristics. Dealing with such characteristics is an attribute of intelligence and autonomy. Uncertainty generally implies that it is necessary to "see to act" (as in control) and to "act to see" (as in perception). It also has strong consequences for the planning process, since appropriate "motion strategies" have to be applied: for instance, how can a blind and inaccurate robot progress towards the exit of a given maze using its own proximity sensors for guiding its motions ? Dealing with complex motion constraints clearly requires devising new models and new search techniques, since the nature of the problem is intrinsically different from traditional path planning which only deals with non-collision constraints and simple integrable kinematic constraints. Even if some interesting results have been recently obtained, motion planning with non-holonomic kinematic constraints and/or dynamic constraints is still an open question. It is a new challenge in robotics and vision to try to solving these problems and integrate, whenever necessary, appropriate planning, sensing and control techniques. A necessary step towards the achievement of such an ambitious goal is to develop *appropriate geometric reasoning techniques* with reasonable computational complexity.

One of the activities of the French Joint Research Programme in Artificial Intelligence (PRC-IA) is to study how geometric reasoning can contribute to solving these problems. A workshop on this topic was held at LIFIA-IRIMAG in Grenoble (France) on September 16-17, 1991. This workshop was jointly organized by LIFIA-IRIMAG and PRC-IA. It was funded by the French Ministry of Research and Space (MRE), the French Ministry of Education (MEN) and LIFIA-IRIMAG. The scientific programme consisted of 18 contributions, 15 of which were selected for inclusion in the present book.

The selected contributions cover several important areas in the field of robotics and computer vision. Part 1 deals with motion planning with kinematic and dynamic constraints. Part 2 investigates motion planning and control in the presence of uncertainty, and presents planning based techniques, reactivity, and visual servoing methods. Part 3 addresses geometric problems related to visual perception. Part 4 deals with numerical problems linked to the implementation of practical algorithms for visual perception.

Impressive results were presented during the workshop in both the fields of motion planning and visual perception. However, there is obviously a lack of joint results in these fields. Hopefully the purpose of several new European joint projects is to study action-perception-control interaction problems. A special issue of the

International Journal of Robotics Research will be dedicated to this topic in the near future.

Finally I would like to thank all the participants who contributed to the success of the workshop and the quality of the present volume.

Grenoble, June 1993

Christian LAUGIER

Table of Contents

Numerical Tools for Visual Perception

Motion Planning
under
Kinematic and Dynamic
Constraints

Shortest Paths of Bounded Curvature in the Plane

Jean-Daniel Boissonnat* and André Cérézo* and Juliette Leblond†

* INRIA Sophia-Antipolis
2004 route des Lucioles
06565 Valbonne, France
† Université de Nice
Parc Valrose
06034 Nice, France

Abstract. Given two oriented points in the plane, we determine and compute the shortest paths of bounded curvature joining them. This problem has been solved recently by Dubins in the no-cusp case, and by Reeds and Shepp otherwise. We propose a new solution based on the minimum principle of Pontryagin. Our approach simplifies the proofs and makes clear the global or local nature of the results.

1 Introduction

The question considered here is the following: given two oriented points (M', θ') and (M'', θ'') in the plane, determine and compute the shortest piecewise regular paths joining them, along which the curvature is everywhere bounded by a given $\frac{1}{R} > 0$. Minimizing the length is meaningful both in the class of paths which are C^1 and piecewise C^2, and in the slightly larger class of paths with a finite number of cusps.

This question appears in many applications; for instance, Markov[3] studied the no-cusp case (of joining pieces of railways). A 3-dimensional version applies to planning plumbary networks, or the version with cusps to any car driver.

Even without obstacles, characterizing the shortest paths is not simple. This was only done recently by Dubins[2] in the no-cusp case, and by Reeds and Shepp[5] otherwise.

Our way of solving the question here is radically different from theirs. It is both much simpler and better adapted to (eventually) generalizations in the case of obstacles limiting, however. The essential tool we use is the powerful result of optimal control theory known as the minimum principle of Pontryagin. We recall in Section 2 its basic version which we will use, and refer to standard books in control theory, like [1], [4] and [7], for its quite direct proof.

In Section 3 we apply the principle to our case, and derive some general crucial lemmas. Section 4 is devoted to the no-cusp case, and Section 5 to the more difficult case with cusps.

Our results are essentially the same as those of [2] and [5]. The interest of the present work lies in the method of proof, both simplified for the use of a

Shortest Paths of Bounded Curvature in the Plane

Jean-Daniel Boissonnat[1] and André Cérézo[2] and Juliette Leblond[1]

[1] INRIA Sophia Antipolis
2004 route des Lucioles
06565 Valbonne, France
[2] Université de Nice
Parc Valrose
06034 Nice, France

Abstract. Given two oriented points in the plane, we determine and compute the shortest paths of bounded curvature joigning them. This problem has been solved recently, by Dubins in the no–cusp case, and by Reeds and Shepp otherwise. We propose a new solution based on the minimum principle of Pontryagin. Our approach simplifies the proofs and makes clear the global or local nature of the results.

1 Introduction

The question considered here is the following : given two oriented points (M_i, θ_i) and (M_f, θ_f) in the plane, determine and compute the shortest piecewise regular paths joining them, along which the curvature is everywhere bounded by a given $\frac{1}{R} > 0$. Minimizing the length is meaningful both in the class of paths which are C^1 and piecewise C^2, and in the slightly larger class of paths admitting a finite number of cusps.

This question appears in many applications : for instance Markov[3] studied the no–cusp case for joining pieces of railways. A 3–dimensional version applies to planning plumbary networks, or the version with cusps to any car driver.

Even without obstacles, characterizing the shortest paths is not simple. This was only done recently, by Dubins[2] in the no–cusp case, and by Reeds and Shepp[5] otherwise.

Our way of solving the question here is entirely different from theirs. It is both much simpler and better adapted to further generalization to the case of obstacles limiting moves. The essential tool we use is the powerful result of optimal control theory known as the "minimum principle of Pontryagin". We recall in Section 2 its basic version which we will use, and refer to classical books in control theory, like [1], [4], and [7], for its quite delicate proof.

In Section 3 we apply the principle to our case, and deduce some general crucial lemmas. Section 4 is devoted to the no–cusp case, and Section 5 to the more difficult case with cusps.

Our results are essentially the same as those of [2] and [5]. The interest of the present work lies in the method of proof, both simplified by the use of a single idea, and as local as the statements will allow. Indeed, we make a clear distinction between local and global proofs, and we insist on local proofs in view of further work dealing with obstacles.

Related results to be reported in a forthcoming publication have been obtained independently by Sussmann and Tang [6]. The results in [2] and [5] are also deduced from the minimum principle and new lights on the piecewise regularity of optimal controls.

2   The minimum principle : a basic version

Given are :

- two integers n and r, two points (x^i) and (x^f) in IR^r, and a compact subset U of IR^n,
- a C^0 function $f(x, u) : IR^n \times U \to IR^n$,
- a C^0 function $f_0(x, u) : IR^n \times U \to IR^n$.

A "control" is a piecewise continuous (but not necessarily continuous) function $u(t) : [0, T] \to U$, for some $T > 0$. $\mathcal{U}$ denotes the set of controls. We want to find a $u \in \mathcal{U}$ which minimizes the integral

$$J(u) = \int_0^T f_0(x(t), u(t)) dt .$$

Here $x(t) = (x_1(t) \cdots, x_n(t)) : [0, T] \to IR^n$ is a solution of the differential system with boundary conditions

$$\begin{cases} \dfrac{dx_j}{dt} = f_j(x, u(t)) \ (j = 1, \ldots, n) \\ x(0) = x^i; \ x(T) = x^f \end{cases} \tag{1}$$

$J(u)$ is called the "cost" of the "path" $x(t)$, solution of (1), given the "control" $u(t)$. The solution $x(t)$ of the system with initial conditions $x(0) = x^i$ is well determined for a given u. This is the case even in a much more general setting (see e.g. [7, Theorem II.4.11]). We denote by $\mathcal{U}_i^f$ the subset of "admissible" controls $u \in \mathcal{U}$ such that the associated paths x satisfy also the final conditions $x(T) = x^f$. The optimal control problem is to find controls $* \in \mathcal{U}_i^f$ satisfying

$$J(*) = \min_{u \in \mathcal{U}^f} J(u)$$

Such an $*$ is called "optimal", as well as the associated path x. Note that T is here arbitrary. But in the particular case $f_0 \equiv 1$, $J(u) = T$ and we want to minimize the "time" to go from x^i to x^f.

We give a more geometric interpretation to the approach by adding an extra variable $x_0(t)$, solution of $\frac{dx_0}{dt} = f_0(x, u(t))$, $x_0(0) = 0$. We are therefore looking for a $u \in \mathcal{U}$ such that the solution of

$$
\begin{cases}
\dfrac{dx_j}{dt} = f_j(x, u(t)) \ (j = 0, \ldots, n) \\[2mm]
x_j(0) = x^i
\end{cases}
\qquad (j = 0, \ldots, n)
\tag{2}
$$

satisfies the conditions $x_j(T) = x_j^f$ $(j = 1, \ldots, n)$ and $J(u) = x_0(T)$ is minimal (for some $T > 0$). A basic idea in mechanics is then to introduce "dual" variables $\psi = (\psi_0, \psi_1, \ldots, \psi_n) : [0, T] \longrightarrow \mathbb{R}^{n+1}$ which are continuous, piecewise C^1, and solution of the "adjoint" system

$$
\begin{cases}
\dfrac{d\psi_j}{dt} = -\displaystyle\sum_{i=0}^{n} \dfrac{\partial f_i}{\partial x_j}(x, u(t))\psi_i \ (j = 0, \ldots, n) \\[4mm]
\psi(0) = \psi^i \, .
\end{cases}
\tag{3}
$$

For a given $u \in \mathcal{U}_i^f$, x the associated solution of (2), and an arbitrary initial condition ψ^i, (3) has a unique solution ψ.

If the "Hamiltonian" is defined by $H(\psi, x, u) : \mathbb{R}^{2n+2+r} \longrightarrow \mathbb{R}$

$$
H(\psi, x, u) = < \psi, f > = \sum_{j=0}^{n} \psi_j f_j(x, u),
\tag{4}
$$

then, equations (2) and (3) can be rewritten as a Hamilton–Jacobi system with parameter u :

$$
\begin{cases}
\dfrac{dx_j}{dt} = \dfrac{\partial H}{\partial \psi_j} \ , \ x_j(0) = x_j^i \\[4mm]
\dfrac{d\psi_j}{dt} = -\dfrac{\partial H}{\partial x_j} \ , \ \psi_j(0) = \psi_j^i \, .
\end{cases}
\qquad j = 0, \ldots, n
\tag{5}
$$

Finally, we define

$$
M(\psi, x) = \min_{u \in U} H(\psi, x, u)
$$

where ψ, x, u are considered as independent variables. The fundamental result of Pontryagin[4] is then :

Theorem 1 (Minimum principle) : *If $*$ is an optimal admissible control, there exists a non-zero adjoint vector ψ, and $T > 0$, such that, $(x(t), \psi(t))$ being the solution of (5) for $u = *$, one has :*

1. *$\forall t \in [0, T]$ $H(\psi(t), x(t), *(t)) \equiv M(\psi(t), x(t))$*
2. *Furthermore, $\forall t \in [0, T]$, $M(\psi(t), x(t)) \equiv 0$ and $\psi_0(t) \equiv \psi_0(0) \geq 0$.*

This changes the question of minimizing the functional $J(u)$ over $\mathcal{U}$ into a minimum problem for the "scalar" function H on U. In some sense, conditions 1 and 2 hereabove insure that $(x(t), \psi(t))$ is "stationary" among the solutions of (5), and the principle asserts that the optimal paths are to be found only among these. The minimum principle of course only gives necessary conditions, and does not even assert that an optimal control exists. Its existence has to be proved independently, and it is usually done in the much larger class of controls which are only assumed to be integrable (see e.g. [7, Theorem V.6.1]). The minimum principle applies just as well in this larger class (see [4, Chapter II]), and the following computations (Section 3) will then prove that the optimal controls in fact belong to our smaller class $\mathcal{U}$. This is why we allow ourselves to restrict our discussion to the class $\mathcal{U}$, as well as for the sake of clarity, and for emphasis on applications.

A new proof of the piecewise regularity of optimal controls can be found in [6], in a general setting which includes our case, based on the properties of subanalytic sets.

Most theorems of existence assume that the range of controls is convex (see e.g. [7, Theorem V.6.1]). This is the technical reason why we consider convex ranges of control in Section 3.

3 Application to shortest paths of bounded curvature

Let us recall that we want to minimize the length of continuous and piecewise C^2 paths $(x(t), y(t))$ in the plane $I\!R^2$ joining given initial and final points with given orientations (x^i, y^i, α^i) and (x^f, y^f, α^f). By assumptions on the control, such paths are formed of a sequence of C^2 paths, glued together at isolated commutation points $t_k \in]0, T[$, where either the curve is C^1 (called "inflexion" points) or the orientation is reversed (called "cusps").

Thus the following quantities are well defined.

- The polar angle $\alpha(t)$ of the tangent to the path, considered as a globally continuous $I\!R/2\pi$-valued function by assuming its continuity at the cusps. Its intuitive meaning is that the tangent to the path is directed as the lights of a car that would follow the path, changing from front to rear gear or conversely at a cusp.

- The arc length along the path, which we denote by t : intuitively, the path is run at constant speed one.

- The curvature $u(t) = \frac{d\alpha}{dt}$, which is defined everywhere except at the commutation points. When $u(t) \neq 0$, its sign depends on whether the point (x, y) runs in clockwise sense $(u < 0)$ or in counter-clockwise sense $(u > 0)$ as t increases.

The differential system (2) can be written as (from now on we write $\dot{z}$ for $\frac{dz}{dt}$) :

$$\begin{cases} \dot{x} = \varepsilon \cos \alpha & x(0) = x^i \quad x(T) = x^f \\ \dot{y} = \varepsilon \sin \alpha & y(0) = y^i \quad y(T) = y^f \\ \dot{\alpha} = u & \alpha(0) = \alpha^i \quad \alpha(T) = \alpha^f \\ \dot{x}_0 = 1 & x_0(0) = 0 \end{cases} \tag{6}$$

with control functions $(\varepsilon, u) \in U \subset I\!R^2$, where $U = \{-1, +1\} \times [-\frac{1}{R}, +\frac{1}{R}]$. This means that we control the instantaneous curvature u, allowing also changes between front and rear gears (the sign of ε). The cost we want to minimize is equal to the length of the trajectory and defined by :

$$J(u, \varepsilon) = \int_0^T dt = T \tag{7}$$

For the technical reasons already mentioned at the end of section 2, we should assume here U convex, more precisely : $U = [-1, +1] \times [-\frac{1}{R}, +\frac{1}{R}]$, that is to say $-1 \le \varepsilon \le 1$. We modify (6) accordingly and consider now the system :

$$\begin{cases} \dot{x} = \varepsilon \cos \alpha & x(0) = x^i \quad x(T) = x^f \\ \dot{y} = \varepsilon \sin \alpha & y(0) = y^i \quad y(T) = y^f \\ \dot{\alpha} = |\varepsilon| u & \alpha(0) = \alpha^i \quad \alpha(T) = \alpha^f \\ \dot{x}_0 = 1 & x_0(0) = 0 . \end{cases} \tag{8}$$

In this case, t is the time, and no longer the arc length s, and we have : $ds = |\varepsilon(t)| \, dt$. The expression (7) of the cost remains unchanged, although it corresponds now to a minimum time problem. Calling (p, q, β, e) a set of dual variables to $(x, y, \alpha, x_0 = t)$, the minimum principle applies here. The Hamiltonian is

$$H = e + \varepsilon p \cos \alpha + \varepsilon q \sin \alpha + |\varepsilon| \beta u , \tag{9}$$

and the adjoint system

$$\begin{cases} \dot{p} = 0 \\ \dot{q} = 0 \\ \dot{\beta} = \varepsilon p \sin \alpha - \varepsilon q \cos \alpha \\ \dot{e} = 0 \end{cases} \tag{10}$$

with arbitrary (but not all zero) initial values. Thus p, q, e are constant on $[0, T]$. Putting $p = \lambda \cos \phi$, $q = \lambda \sin \phi$, with $\lambda = \sqrt{p^2 + q^2} \ge 0$, determines an angle ϕ modulo 2π, such that $\tan \phi = \frac{q}{p}$. We rewrite (9) and (10) as :

$$H = e + \varepsilon \lambda \cos(\alpha - \phi) + |\varepsilon| \beta u \tag{11}$$

$$\dot{\beta} = \varepsilon \lambda \sin(\alpha - \phi) \tag{12}$$

As H is a piecewise affine function in ε, it cannot reach its minimum w.r.t. ε elsewhere than at $\varepsilon = 0, \pm 1$. But $\varepsilon \equiv 0$ on some interval is obviously irrelevant since it corresponds to zero velocity which is trivially not an optimal control for a minimum time problem. Thus, condition 1 of the minimum principle asserts that optimal controls can only be obtained for $\varepsilon = \pm 1$, so that we set $|\varepsilon| = 1$ in the following of the discussion. Hence, the arc length becomes equal to t, and minimum time solutions provide minimum length paths. Moreover, H rewrites

$$H = e + \varepsilon\lambda \cos(\alpha - \phi) + \beta u \qquad (13)$$

and systems (6) and (8) are identical. Condition 1 also states that for an optimal control (u, ε), along any C^2 piece of the optimal path, we have

$$\lambda\varepsilon \cos(\alpha - \phi) \leq 0 \ and \ \beta u \leq 0. \qquad (14)$$

Furthermore, either one of the following two cases holds :

- $\frac{\partial H}{\partial u} = \beta \equiv 0$, thus $\dot{\beta} \equiv 0$, and $\alpha \equiv \phi$ or $\alpha \equiv \phi + \pi$, and the path is a line segment with direction ϕ (notice that $\lambda \neq 0$, otherwise, the adjoint vector will be 0, which is forbidden by the minimum principle),
- or $\frac{\partial H}{\partial u} \not\equiv 0$ and thus, by condition 1, $u = \pm\frac{1}{R}$, and the path is an arc of circle of radius R.

The preceding discussion leads immediately to :

Proposition 2 *Any optimal path is the concatenation of arcs of circles of radius R, and line segments all parallel to some fixed direction ϕ.*

Also condition 2 of the minimum principle ($H \equiv 0$ and $e \geq 0$) implies :

Lemma 3 $\beta = 0$ *at the inflexion points and on the line segments.*

Proof : That $\beta = 0$ on the line segments was already mentioned. It implies $\beta = 0$ at the inflexions between a line segment and an arc of circle, since β is continuous. But at an inflexion between two arcs, u changes sign, and so should β in order that $\beta u \leq 0$; thus the same continuity argument applies. $\square$

Lemma 4 *If $\lambda = 0$, the optimal path is either a line segment, or a sequence of arcs of radius R joined by cusps.*

Proof : (13) and condition 2 imply $\beta u \equiv -e$. If $e = 0$, β cannot vanish since $\psi = (p, q, \beta, e)$ would then vanish, which is forbidden by the principle. Thus $u \equiv 0$, and the whole path is one line segment. If $e > 0$, then neither β nor u can vanish, and the path contains neither line segments nor inflexions, by Lemma 3. $\square$

In the following, we assume $\lambda \neq 0$.

Lemma 5 $e > 0$

Proof : Assume that $e = 0$. Then, by (13), $\beta u + \varepsilon\lambda \cos(\alpha - \phi) \equiv 0$. But the two parts of the sum are of the same sign by (14) so that they have to be zero. Since $|\varepsilon| = 1$ and $\lambda \neq 0$, this implies (i) : $\cos(\alpha - \phi) = 0$ and (ii) : $\beta u = 0$. By (i), $\alpha = \phi \pm \frac{\pi}{2}$ is constant, and this path is a line segment along which, necessarily, $\beta = u = 0$ and $\dot\beta = 0$. This implies $\alpha = \phi$ or $\alpha = \phi + \pi$, and leads to a contradiction. Hence $e \neq 0$ and the result follows from the condition 2 ($e \geq 0$) of Theorem 1. $\square$

Lemma 6 $\beta - py + qx$ *is constant along any optimal path. Consequently, all the points of an optimal path where β takes the same value are on the same straight line of direction ϕ.*

Proof : According to (6) and (10) $\dot\beta - p\dot y + q\dot x \equiv 0$, and thus $\beta - py + qx \equiv c$ for some $c \in IR$, on the whole optimal path. $\square$

Lemma 6 has the two following consequences :

Lemma 7 *On any optimal path, the line segments and the inflexion points are all on a straight line D_0 with direction ϕ (of equation $py - qx + c = 0$).*

Proof : Apply Lemma 6 with $\beta = 0$ $\square$

Lemma 8 *All the cusps of any optimal path are on two straight lines $D_\pm$ parallel and equidistant to D_0, of equation $py - qx + c = \mp e R$. Furthermore, (a) there are no cusps between a line segment and an arc of circle, (b) the "positive" cusps (that is where $u > 0$ both before and after) are on D_+, and the others on D_-, (c) the half-tangent at a cusp is perpendicular to $D_\pm$.*

Proof : By condition 2 of the principle, $0 \equiv e + \varepsilon\lambda \cos(\alpha - \phi) + \beta u$. Near a cusp, u remains constant and equal to $\pm\frac{1}{R}$, while ε changes from $+1$ or -1 or the converse. Hence, by (14), $\cos(\alpha - \phi) = 0$; thus $\alpha = \phi \pm \frac{\pi}{2}$ and $\beta = \pm e R$. The first assertion follows from Lemma 6, then (a) and (c) from Lemma 7, and (b) from (14). $\square$
Moreover, we have :

Lemma 9 *Any C^2 arc of circle of an optimal path, between two points where $\beta = 0$, has length $> \pi R$.*

Proof : Assume the length of such an arc, between $t = t_1$ and $t = t_2$, to be $\leq \pi R$. As $\beta(t_1) = \beta(t_2) = 0$, we have by (13) and since $H \equiv 0$,

$$\varepsilon \cos(\alpha(t_1) - \phi) = \varepsilon \cos(\alpha(t_2) - \phi) = -\frac{e}{\lambda} \tag{15}$$

β is C^1 on $[t_1, t_2]$ and it keeps a constant sign on an arc, by (14). So, β reaches an extremum at some point $t_3 \in]t_1, t_2[$ where $\dot\beta(t_3) = \varepsilon\lambda \sin(\alpha(t_3) - \phi) = 0$ by (12). Now, $\varepsilon \cos(\alpha(t_3) - \phi) = -1$, by (15) and since the sign of $\cos(\alpha(\dot t_3) - \phi)$ cannot change on $[t_1, t_2]$, according to our assumption on the length. But then

$$0 \equiv H = e - \lambda + \beta(t_3)\, u(t_3) \leq e - \lambda,$$

by (14), and thus $\frac{e}{\lambda} \geq 1$, which compared with (15) implies $e = \lambda$. Since $|*(t_3)| = \frac{1}{R} > 0$, $\beta(t_3) = 0$. Hence $\beta \equiv 0$ on $[t_1, t_2]$; this means by (12) that this piece of path would be a line segment, not an arc of circle. □

We remark now that we may simplify our equations by using obvious symetries. The rectangular coordinate system in IR^2 is arbitrary. So, we can assume $\phi = 0$ by a rotation of axes, $\lambda = 1$ or 0 since the change from $\frac{\psi}{\lambda}$ to ψ leaves the minimum principle invariant when $\lambda > 0$, and the constant c in Lemmas 6, 8 to be zero by a translation of the origin in IR^2. Under these assumptions we have $p = 1$, $q = 0$. The lines D_0, $D_\pm$ have now the equations $y = 0$ and $y = \mp e R$, and $\beta \equiv y$ is the signed distance of a point on the path to D_0. Furthermore, we have

$$H = e + \varepsilon \cos \alpha + u \beta \equiv 0. \tag{16}$$

Consider now an arc of circle with a cusp at one of its endpoints. If this endpoint is on D_-, $u = -\frac{1}{R}$ and, since at a cusp $\alpha = \pm\frac{\pi}{2}$, by (16), $0 \le \beta = e R + \varepsilon R \cos \alpha \le e R$ since $\varepsilon \cos \alpha \le 0$ by (14). The same argument shows that on an arc with a cusp on D_+, one has $0 \ge \beta \ge -e R$. Summarizing this discussion, we get :

Proposition 10 *For any optimal path with $\lambda \ne 0$, there exists a rectangular coordinate system in IR^2 such that (see figure 1)*

- *all the line segments and inflexion points on the path lie on the first axis D_0 : $y = 0$,*
- *all the cusps where $u > 0$ lie on the line D_+ : $y = -e R$,*
- *all the cusps where $u < 0$ lie on the line D_- : $y = +e R$,*
- *an arc of circle with a cusp as one of its endpoints on D_+ (D_-) lies between D_+ and D_0 (between D_0 and D_-),*
- *(16) is satisfied at any point on the path.*

Remark 1 *(a) Outside the three lines D_0, D_+, D_- there can be no commutation and no line segment. A piece of an optimal path not intersecting these lines can only consist of one arc of circle, and is contained in one of the strips $[D_+, D_0]$ and $[D_0, D_-]$ as soon as it is not an initial or a final arc.*
(b) By (16) and since $u \ne 0$ outside D_0, the geometric angle of the tangent line to an optimal path with D_0 at a point (x, y) is almost determined by the "distance" y of the point to D_0. Indeed α satisfies

$$\varepsilon \cos \alpha = \pm\frac{y}{R} - e \le 0 \tag{17}$$

which gives only two possible values of α for any piece of path with no commutation (that is for given ε and u).
(c) As soon as an optimal path contains a line segment, the tangent at the inflexion points is D_0 itself. Indeed, at a point on a line segment, $y = 0$, $\alpha = 0 \bmod \pi$ by Proposition 2, thus by (16) $e = \pm\varepsilon = 1$, and (17) then gives $\cos \alpha = \pm 1$ at any point on D_0.

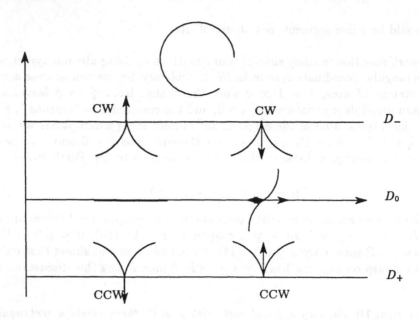

Fig. 1. : D_0, D_+, D_-

Proposition 10 and Remark 1 invite us to classify the different possible kinds of optimal paths according to whether $e > 1$, $e = 1$, or $0 < e < 1$. We will denote each concatenation of segments and arcs by a word like for instance "$C|CSC|C$", each C meaning an arc of circle of radius R, S a line segment of direction $\phi = 0$, and | a cusp ; C_v will mean an arc of circle of length Rv.

4 The C^1 (no cusp) case

The preceding study applies to the characterization of C^1 shortest paths between given initial and final positions and orientations in the free plane. It is enough to set ε to be 1 and allow no cusps in all the statements above (but here $\cos\alpha$ can take arbitrary sign as we do not minimize anymore with respect to ε).

As soon as an optimal path contains a line segment, it has to be of CSC type (or the degenerate forms CS, SC, S). Indeed, since S is on D_0, any other event would mean another commutation on D_0, thus a full circle, which is obviously not optimal.

That an optimal path without segment is necessarily of the CCC kind (or the degenerate forms CC or C) is a consequence of the remark that a portion of an optimal path is itself an optimal path, and from the following lemma :

Lemma 11 *No path of type CCCC is optimal.*

We give a direct local proof of this lemma, by showing that there is always a strictly shorter path arbitrarily close to the initial one, having the same endpoints, and of the same type $CCCC$.

Proof : The notations being clear on Figure 2, let us write ν_θ for the unitary vector of polar angle θ, and assume $R = 1$ for simplicity.
If such a path of the $CCCC$ kind were optimal for some values of a, b and for

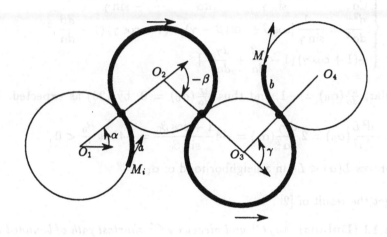

Fig. 2. : The case $CCCC$

$\alpha = \alpha_0$, $\beta = \beta_0$, $\gamma = \gamma_0$, Lemmas 7 and 9 would imply $\beta_0 = \gamma_0$ and that the intermediary arcs are each of length $> \pi$. Thus

$$\begin{cases} O_4 = O_1 + 2\nu_{\alpha_0} + 2\nu_{\alpha_0-\beta_0} + 2\nu_{\alpha_0-\beta_0+\gamma_0} \\ L_0 = a + (\pi + \beta_0) + (\pi + \gamma_0) + b \end{cases} \tag{18}$$

where L_0 is the length of the path. We can deform this path into a similar one with the same endpoints but slightly different angles α, β, γ, and we get in the same way (noting L its length)

$$\begin{cases} O_4 = O_1 + 2\nu_\alpha + 2\nu_{\alpha-\beta} + 2\nu_{\alpha-\beta+\gamma} \\ L = a + (\alpha - \alpha_0) + (\pi + \beta_0) + (\pi + \gamma_0) + b \\ +(\alpha - \beta + \gamma - \alpha_0 + \beta_0 - \gamma_0) \end{cases} \tag{19}$$

so that

$$L - L_0 = 2(\alpha - \alpha_0) + 2(\gamma - \gamma_0), \; thus \; dL = 2(d\alpha + d\gamma) \tag{20}$$

$$\begin{cases} \cos\alpha + \cos(\alpha - \beta) + \cos(\alpha - \beta + \gamma) \\ = \cos\alpha_0 + \cos(\alpha_0 - \beta_0) + \cos(\alpha_0 - \beta_0 + \gamma_0) \\ \sin\alpha + \sin(\alpha - \beta) + \sin(\alpha - \beta + \gamma) \\ = \sin\alpha_0 + \sin(\alpha_0 - \beta_0) + \sin(\alpha_0 - \beta_0 + \gamma_0) \end{cases} \quad (21)$$

The Jacobian of (21) with respect to β and γ is equal to $-\sin\gamma$ thus not zero near $(\alpha_0, \beta_0, \gamma_0)$ since $0 < \beta_0 = \gamma_0 < \pi$. So system (21) defines implicitly β and γ as functions of α in a neighborhood of $(\alpha_0, \beta_0, \gamma_0)$, which proves that such a deformation is possible. Differentiating twice (21) with respect to α, we get

$$\begin{cases} \dfrac{d\beta}{d\alpha} = \dfrac{\sin(\beta - \gamma) - \sin\gamma}{-\sin\gamma} \;,\; \dfrac{d\gamma}{d\alpha} = \dfrac{\sin(\beta - \gamma) + \sin\beta}{-\sin\gamma} \\ \dfrac{d^2\gamma}{d\alpha^2} = \dfrac{-1}{\sin\gamma}\left[\cos\beta + \cos(\beta - \gamma) + (1 + \cos\gamma)\left(1 - \dfrac{d\beta}{d\alpha}\right)^2\right. \\ \left. + (1 + \cos\gamma)\left(1 - \dfrac{d\beta}{d\alpha} + \dfrac{d\gamma}{d\alpha}\right)^2\right] \end{cases}$$

In particular, $\frac{d\gamma}{d\alpha}(\alpha_0) = -1$, and thus $\frac{dL}{d\alpha}(\alpha_0) = 0$, by (20) as expected. But further

$$\frac{d^2 L}{d\alpha^2}(\alpha_0) = 2\frac{d^2\gamma}{d\alpha^2}(\alpha_0) = -4\frac{1 + \cos\beta_0}{\sin\beta_0} = -4\cot\frac{\beta_0}{2} < 0 \,,$$

and this proves $L(\alpha) < L_0$ in a neighborhood of α_0. $\qquad\square$

Thus we get the result of [2] :

Theorem 12 (Dubins) *Any C^1 and piecewise C^2 shortest path of bounded curvature in the free plane between given endpoints and orientations is either of type CSC, or of type CC_vC with $v > \pi$, or a degenerate form of these.*

Remark 2 *Dubins[2] (and Reeds and Shepp[5] in case of cusps) seem to get a more general result, as they minimize the length in a larger class of paths. But our conclusion amounts to the same, as we already explained at the end of Section 2.*

5   The general case (allowing cusps)

We go back to the general statements and results of Section 3. The case of an optimal path for which $\lambda = 0$ has already been settled by lemma 4. Notice that in this case, any path of type S is obviously optimal, as well as any path of type $C_{v_1}|C_{v_2}|\ldots|C_{v_p}|\ldots$ when $v_1 + \ldots + v_p + \ldots \leq \pi$, since all these paths have the same length $R(v_1 + \ldots + v_p + \ldots) = R|\alpha^f - \alpha^i|$; but the number of cusps is unbounded and may even be infinite, since we can always replace a sequence $C_{v_1}|C_{v_2}|C_{v_3}$ by $C_{v_1'}|C_v|C_{v_2'}|C_w|C_{v_2'}$ with $v_1 + v_2 + v_3 = v_1' + v + v_3' + w + v_2'$. It can be shown that the converse is also true. Thus there are in this case infinitely many shortest paths, in striking contrast to section 4. Notice however that there always exists one with at most two cusps, as proved by [5].

In the following we focus on the case of an optimal path for which $\lambda \neq 0$, where the whole of Section 3 applies, and in particular Proposition 10 (with $\lambda = 1$, $\phi = c = 0$). We assume that there is at least one cusp on the path (the no–cusp case being already settled) and we lead the discussion according to the values of $e > 0$.

5.1 $e > 1$

No arc of circle of radius R perpendicular to $D_\pm$ can reach D_0. Thus, there is no inflexion, and the path is of a type $C|(C_\pi|)^k\,C$ for some $k \geq 0$.

5.2 $0 < e < 1$

Any circle of radius R centered on $D_\pm$ intersects D_0 at angles satisfying $\cos \alpha = -\varepsilon e \neq 0$. Thus the possible values for α are $\pm v, \pm v + \pi$ for some $0 < v < \frac{\pi}{2}$.

But then part (c) of Remark 1 asserts that the path contains no line segment and an arc of circle starting at a cusp A, on D_+ (D_-), can only

1. either be final : $|C$,
2. or end at an inflexion point I_1 : $|C_v C \ldots$ or I_2 : $|C_{\pi-v} C \ldots$,
3. or end at a cusp : $|C_\pi|$.

An inflexion at I_2 (the inflexion point farthest from A) is impossible, since the arc $I_1 I_2$ would be an arc between two points where $\beta = y = 0$ and shorter than πR, contrary to Lemma 9. The third case is also impossible by the same argument since the arc intersects D_0 in two points. Thus we are left with only $|C$ or $|C_v C$, the last arc being again either final or ending at a new cusp on D_- (D_+), and thus also of length vR.

We conclude that any portion of an optimal path starting at a cusp is of one of the three following types $|C$, $|C_v C$ or $|C_v C_v| \ldots$ with $0 < v < \frac{\pi}{2}$, and the same v along the whole path.

Hence the possible types of optimal paths in the case $0 < e < 1$ are

$$C|C, \ C|C_v C, \ CC_v|C, \ CC_v|C_v C, \ C|C_v C_v|C$$

with $0 < v < \frac{\pi}{2}$ (compare to the list in [5]), and also others of type

$$\ldots CC_v|C_v C_v|C \ldots, \ or \ \ldots C|C_v C_v|C_v C \ldots$$

The following lemma shows that the two last cases cannot be part of an optimal path by computing a local deformation of the paths, similarly to the proof of Lemma 11.

Lemma 13 *No path of type $CC_v|C_v C_v|C$ (or $C|C_v C_v|C_v C$) with $0 < v < \frac{\pi}{2}$ can be optimal.*

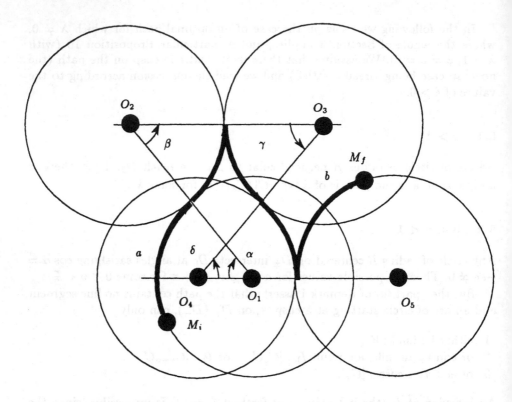

Fig. 3. : The $CC_v|C_vC_v|C$ case

Proof : We consider only the case $CC_v|C_vC_v|C$, the computations for the other one being identical. The notations being clear on Figure 3, let again ν_θ denote the unitary vector with polar angle θ and L the length of the path. Assume $R = 1$ for simplicity.

For any nearby path of the same type and with the same endpoints $CC_\beta|C_\gamma C_\delta|C$, O_1 and O_5 are unchanged and we can take O_1O_5 as the x-axis, so that

$$\begin{cases} O_5 = O_1 + 2\nu_{\pi-\alpha} + 2\nu_{-\alpha+\beta} + 2\nu_{\pi-\alpha+\beta+\gamma} + 2\nu_{-\alpha+\beta+\gamma-\delta} \\ L = a + (\alpha - v) + \beta + \gamma + \delta + b - (\alpha - \beta - \gamma + \delta) \end{cases}$$

while for the initial path $\alpha = \beta = \gamma = \delta = v$ by Proposition 10. Hence

$$\begin{cases} O_5 = O_1 + 2\nu_{\pi-v} + 2\nu_0 + 2\nu_{\pi+v} + 2\nu_0 \\ L_0 = a + v + v + v + b \quad \textit{for some positive } a, b \end{cases}$$

This yields

$$L - L_0 = 2(\beta - v) + 2(\gamma - v), \textit{ thus } dL = 2d\beta + 2d\gamma \tag{22}$$

$$\begin{cases} -\cos\alpha + \cos(\beta - \alpha) - \cos(\beta + \gamma - \alpha) + \cos(\beta + \gamma - \alpha - \delta) \\ = 2 - 2\cos v \\ \sin\alpha + \sin(\beta - \alpha) - \sin(\beta + \gamma - \alpha) + \sin(\beta + \gamma - \alpha - \delta) = 0 \end{cases} \tag{23}$$

Now we look for shorter paths assuming furthermore $\delta = \gamma$. Equations (22) and (23) become

$$L - L_0 = 2(\beta + \gamma - 2v) \tag{24}$$

$$\begin{cases} -\cos\alpha + 2\cos(\beta - \alpha) - \cos(\beta - \alpha + \gamma) = 2 - 2\cos v \\ \sin\alpha + 2\sin(\beta - \alpha) - \sin(\beta - \alpha + \gamma) = 0 \end{cases} \tag{25}$$

The Jacobian of (25) with respect to β and γ is equal to $-2\sin\gamma \neq 0$ for γ sufficiently close to v, since $0 < v < \frac{\pi}{2}$. So (25) defines implicitly β and γ as functions of α in a neighborbood of (v, v, v). This proves that such a deformation is possible, and differentiating (25) w.r.t. α yields

$$\frac{d\beta}{d\alpha} = 1 - \frac{\sin(\beta + \gamma)}{2\sin\gamma} \quad and \quad \frac{d\gamma}{d\alpha} = \frac{2\sin\beta - \sin(\beta + \gamma)}{-2\sin\gamma}$$

In particular,

$$\frac{dL}{d\alpha} = 2\left(\frac{d\beta}{d\alpha} + \frac{d\gamma}{d\alpha}\right) = 0 \; at \; \beta = \gamma = v.$$

But further,

$$\frac{d^2L}{d\alpha^2}(0) = -4\frac{\cos v}{\sin^2 v}(1 - \cos v) < 0$$

which proves $L < L_0$ for any small enough $\alpha \neq 0$. $\qquad\square$

5.3 $e = 1$

Any circle of radius R centered on $D_\pm$ is tangent to D_0. Thus any arc of an optimal path starting at a cusp, say A on D_+

- either is final : $|C$,
- or ends at a cusp on D_+ : $|C_\pi|$,
- or has length $\frac{\pi}{2}R$ and ends at an inflexion point B, and is followed by
  - either a segment BB' of D_0. If segment BB' is not terminal, then B' is another inflexion point, followed by an arc of circle which, again, is either terminal or of length $\frac{\pi}{2}R$ and ends at a cusp on either D_+ or D_-.
  - or a final arc of circle,
  - or another $C_{\pi/2}$ ending at a cusp A' on D_-.

We conclude that any portion of an optimal path starting at a cusp is of one of the four following types $|C$, $|C_{\pi/2}SC$, or $|C_{\pi/2}SC_{\pi/2}|\ldots$ and its degenerate cases $|C_\pi|\ldots$ and $|C_{\pi/2}C_{\pi/2}|\ldots$

Hence any optimal path, in the case $e = 1$, is of one of the following types (together with their degenerate forms) :

$$C|C, C|C_{\pi/2}SC, CSC_{\pi/2}|C, C|C_{\pi/2}SC_{\pi/2}|C$$

(compare to the list in [5]), plus others of type

$$...CSC_{\pi/2}|C_{\pi/2}SC... ,$$

for instance

$$...C_w SC_{\pi/2}|C_{\pi/2}SC_{\pi/2}|..., \quad ...|C_{\pi/2}SC_{\pi/2}|C_{\pi/2}SC_w ...$$

Summarizing the above discussion, we state :

Theorem 14 *Any shortest path in the plane, piecewise C^2, and either C^1 or with cusps at junction points, between two given points where the orientations are also specified, is of one of the types listed below, together with their degenerate forms :*

CSC, CC_vC (with $v > \pi$),
$C|C|...|C, \ C|C_vC, CC_v|C, CC_v|C_vC, \ C|C_vC_v|C$ (with $0 < v < \frac{\pi}{2}$),
$CSC_{\pi/2}|C_{\pi/2}SC, \ C|(C_{\pi/2}SC_{\pi/2}|)^kC$, (with $k \geq 0$)

It is worth noticing that all the arguments used so far were of a local nature. In the sequel, we will use global arguments and further restrict the number of possible types of shortest paths.

Lemma 15 *A path of type $CSC_{\pi/2}|C_{\pi/2}SC_{\pi/2}|C$ (or $C|C_{\pi/2}SC_{\pi/2}|C_{\pi/2}SC$) cannot be optimal.*

Proof : We consider only the case $CSC_{\pi/2}|C_{\pi/2}SC_{\pi/2}|C$, the computations for the other one being identical. The four possible cases of type $CSC_{\pi/2}|C_{\pi/2}SC_{\pi/2}|C$ (see (1) to (4), figure 4) can be discarded by global arguments. We apply the following length preserving transformation to any of the paths (1) to (4). First, if the two cusps are not already on the same line (case (2) and (4)), reverse a $C_{\pi/2}|C_{\pi/2}$ portion, in order to get the two cusps on the same line, say D_+. Now, transform $SC_{\pi/2}|C_{\pi/2}SC_{\pi/2}|$ into $SC_{\pi/2}|C_\pi|$, by sliding a cusp along D_+. Finally, reverse $|C_\pi|$ in order to remove the cusp. By this transformation, we obtain a path without cusp, which is of the same length than the initial one, and which contains a $SC_{\pi/2}C_\pi$ section. Then, Lemma 9 applies to prove that such a path is not optimal. $\quad\quad\square$

Finally, observe that a path of type CC_vC with $v > \pi$ is no longer optimal. Indeed, by reversing C_v, one obtains a path of type $C|C_w|C$ where $w = 2\pi - v < \pi$, so that it is shorter than the initial path.

We conclude with the following theorem :

Theorem 16 (Reeds and Shepp) *Any shortest path in the plane, piecewise C^2, and either C^1 or with cusps at junction points, between two given points where the orientations are also specified, is of one of the types listed below, together with their degenerate forms :*

$CSC,$
$C|C|...|C, \ C|C_vC, CC_v|C, CC_v|C_vC, \ C|C_vC_v|C$ (with $0 < v < \frac{\pi}{2}$),
$CSC_{\pi/2}|C_{\pi/2}SC, \ C|C_{\pi/2}SC_{\pi/2}|C$

(1)

(2)

(3)

(4)

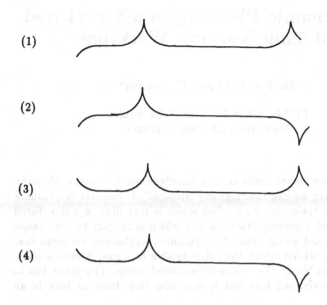

Fig. 4. : The $CSC_{\pi/2}|C_{\pi/2}SC_{\pi/2}|C$ case

Furthermore, the only two cases where there is an infinity of shortest paths are $C|C|\ldots|C$ and $CSC_{\pi/2}|C_{\pi/2}SC$. But one of them can always be found of the type $C|C|C$, or $C|C_{\pi/2}SC$ (or $CSC_{\pi/2}|C$) respectively. This is Reeds and Shepp's result in [5].

Furthermore, in free space, it cannot be optimal unless $k = 0$, since reversing a $|C_\pi|$ arc would produce a path of equal length which cannot be optimal by Lemma 9.

References

1. V.M. Alekseev, V.M. Tikhomirov, and S.V. Fomin. *Optimal control.* Plenum publishing corporation, 1987.
2. L.E. Dubins. On curves of minimal length with a constraint on average curvature and with prescribed initial and terminal positions and tangents. *American Journal of Mathematics*, 79:497–516, 1957.
3. M.G. Krein and A.A. Nudel'man. *The Markov moment problem and extremal problems.* The American Mathematical Society, 1977.
4. L.S. Pontryagin, V.R. Boltyanskii, R.V. Gamkrelidze, and E.F. Mishchenko. *The Mathematical Theory of Optimal Processes.* Interscience, 1962.
5. J.A. Reeds and L.A. Shepp. Optimal paths for a car that goes both forwards and backwards. *Pacific Journal of Mathematics*, 145(2), 1990.
6. H.J. Sussmann and G. Tang. Shortest paths for the Reeds–Shepp car : a worked out example of the use of geometric techniques in nonlinear optimal control. Technical Report 91-10, SYCON, 1991.
7. J. Warga. *Optimal Control of differential and functional equations.* Academic Press, 1972.

Kinodynamic Planning in a Structured and Time-Varying Workspace

Th. Fraichard and C. Laugier*

LIFIA-IRIMAG, 46, av. Félix Viallet
38031 Grenoble Cedex, France

Abstract. This paper deals with a kinodynamic trajectory planning problem [5] that we call the 'highway problem'. It consists in planning a time-optimal trajectory for a robot which is travelling in a structured workspace amidst moving obstacles and which is subject to constraints on its velocity and acceleration. By structured workspace, we mean that there are lanes within which the robot is able to move. A lane is characterized by its 'spine', i.e. a one-dimensional curve. The robot has to follow a predetermined lane but it may also shift from its lane to an adjacent one.

This paper presents an efficient method which determines an approximate time-optimal solution to the highway problem. The approach consists in discretizing time and selecting the acceleration applied to the robot among a discrete set. These hypotheses make it possible to define a grid in the robot's state-time space, i.e. the robot's state (or phase) space augmented of the time dimension. This grid is then searched in order to find a solution. Accordingly trajectory planning is reduced to graph search.

The choice of the time-step determines the size of the grid which, in turn, determines the average running time of the algorithm as well as how close the approximate solution is to the exact time-optimal solution. Thus it is possible to trade off the computation speed against the quality of the solution.

1 Introduction

1.1 Overview of the Problem

Planning motions for robots is a fundamental problem in Robotics which encompasses a wide range of approaches and assumptions. However it seems important to make a distinction between *path planning* and *trajectory planning*. Path planning is characterized by the search of a continuous sequence of collision-free configurations[1] between a start and a goal configuration whereas trajectory planning is concerned with the time history of such a sequence. A *trajectory* is a continuous function of time specifying the robot's configuration at each instant.

---

* Research director at INRIA.

[1] The *configuration* of a robot is a set of independent parameters of minimal cardinality which uniquely defines the position and orientation of every point of the robot.

Trajectory planning is an extension of path planning, the introduction of time permits to consider a wider range of motion planning problems. Thus it is possible to plan motions avoiding moving obstacles or to take into account *dynamic* constraints, i.e. constraints on the velocities attained, the accelerations allowed and the forces applied to the robot. Works on motion planning have traditionally focused on path planning with the 'piano mover' paradigm (we refer the reader to [16] for a recent survey of this topic). Since time is not taken into account in path planning, the goal is generally to find motions which minimize the distance travelled by the robot. In the framework of trajectory planning, it is much more natural to minimize the time it takes to perform a motion, i.e. to look for a *time-optimal* trajectory.

This paper deals with a particular trajectory planning problem that we call the 'highway problem'. It consists in planning a time-optimal trajectory for a robot which is travelling in a structured workspace amidst moving obstacles and is subject to constraints on its velocity and acceleration. By structured workspace, we mean that there are lanes within which the robot is able to move. The robot has to follow a predetermined lane but it may also shift from its lane to an adjacent one. A lane is characterized by its 'spine', i.e. a one-dimensional curve that a robot must follow in a given direction. Such lanes may be a priori defined by the intrinsic structure of the workspace (as in [10] or [25]) but they may also be automatically extracted from a description of the workspace (as in [2] or [3]).

Planning the motion of a car on the highway among other cars is a vivid example of this kind of problem[2] hence the name 'highway problem'. The term 'highway' strongly evokes the idea of a two-dimensional workspace but the highway problem may also refer to motion planning in a three-dimensional workspace such as the airspace which is structured into airways. However the rest of this paper will focus on the two-dimensional case.

A straightforward application of this research work can be found in the European project 'PROMETHEUS Pro-Art' whose purpose is to design an intelligent co-pilot for a car. This framework led us to take into account another constraint: in a workspace such as the roadway, there are several potential moving obstacles (cars, pedestrians, etc.) and it is impossible to have a full a priori knowledge of their motions. Therefore we will assume that the knowledge that we have of the motions of the moving obstacles is restricted to a certain time interval — the *time-horizon*. This time-horizon represents the duration over which an estimation of the motions of the moving obstacles is sound. The main consequence of this assumption is to define an upper bound on the time available to plan the motion of the robot considered.

---

[2] Indeed the basic behaviour of a car is either to follow its current lane while adjusting its velocity (so as to avoid collision with the car in front) or to shift from its lane to an adjacent one (so as to reach an exit or to overtake).

1.2 Contribution of the Paper

This paper presents an efficient method which solves the highway problem. The complexity results which are presented in §2.1 indicate that this problem is an intricate one. Therefore, in order to meet the time-horizon constraint, our method determines an approximate time-optimal solution. This solution is approximate because the search for the solution is restricted to a particular class of trajectories. We find out the time-optimal trajectory in the class considered but it is not necessarily the exact time-optimal solution.

Our approach was initially motivated by the method presented in [5] and which consists in discretizing time and selecting the accelerations applied to the robot among a discrete set. These hypotheses make it possible to define a grid in the robot's state (or phase) space. This grid is then searched in order to find a solution. Accordingly trajectory planning is reduced to graph search. However, in our case, moving obstacles have to be dealt with, therefore we will define the grid in the *state-time space* of the robot, i.e. its state space augmented of the time dimension.

As we will see further down. the choice of the time-step determines the size of the grid which, in turn, determines the average running time of the algorithm as well as how close the solution is to the exact time-optimal solution. Thus it is possible to trade off the computation speed against the quality of the solution.

The paper is organized as follows: §2 briefly reviews the complexity issues and the works related to trajectory planning. Then §3 formally states the highway problem. This statement relies on the specific structure of the workspace considered. §4 describes the algorithm developed in order to solve this problem while §5 presents an implementation of this algorithm along with experimental results.

2 Complexity Issues and Related Works

2.1 Complexity of the Problem

The 'piano mover' problem, i.e path planning, is well-known for its complexity. Various instances of this problem have been shown to be NP-hard, P-space-hard or P-space-complete (see [16]). In general, the time required to solve a path planning problem quickly increases with the dimension of the configuration space considered and the 'complexity' of the obstacles cluttering the workspace.

Since trajectory planning takes into account an extra dimension — time — and extra dynamic constraints, one can expect it to be as computationally expensive as path planning, and it is indeed. For instance, [18] describes a polynomial-time exact algorithm which plans the motion of a point with bounded acceleration amidst moving obstacles in a one-dimensional workspace but [5] shows that the case of a point with bounded velocity and moving in a plane among translating convex polygons is NP-hard. As for a three-dimensional workspace, planning the motion of a disk amidst rotating obstacles is P-space-hard if the velocity of the disk is bounded and NP-hard otherwise [20]. As regards planning the motion of a point subject to bounds on its acceleration and velocity amongst static obstacles, there is an exact algorithm in the two-dimensional case which runs in P-space and takes exponential time [6], but the three-dimensional case is known to be NP-hard [7].

These results suggest that solving the two-dimensional instance of the highway problem is computationally involved. Therefore, in order to have a chance to meet the time-horizon constraint, a few simplifying assumptions are required. These assumptions rely upon the specific structure of the workspace considered. They are presented in §3 along with the formal statement of the problem.

2.2 Related Works

There is a large body of works related to trajectory planning. Some of them focus on trajectory planning amidst moving obstacles while others deal more particularly with dynamic constraints.

Moving Obstacles. The presence of moving obstacles in the workspace of a robot is one reason which makes it necessary to plan a trajectory rather than a path. Indeed a configuration which is collision-free at time t may be occupied by a moving obstacle at time t'.

A general approach developed in order to deal with moving obstacles is to add the time dimension to the robot's configuration space. Thus a new space is obtained which is called *configuration-time space*. A point in this configuration-time space represents a configuration of the robot at a certain time. A moving obstacle maps into a static region of the configuration-time space of the robot. Therefore it is possible to reduce motion planning for a robot amidst moving obstacles to path planning for a point in a static space. Accordingly the different approaches which have been developed in order to solve the path planning problem can be used (see [16]). However these approaches must be adapted in order to take into account a major specificity of the time dimension, i.e. the fact that it is impossible to move backward in time! Among the various existing works, we can distinguish those based upon extensions of the concept of 'visibility graph' [9, 20, 13] and those based upon a cell decomposition of the configuration-time space [12, 16, 22].

An alternate approach which deals with moving obstacles is the so-called 'path-velocity decomposition'. The basic idea is to decompose the trajectory planning into two sub-problems: (a) planning a path which avoids collision with the static obstacles of the workspace and (b) planning the velocity along this path in order to avoid collision with the moving obstacles. This approach was first introduced in [15]. It is efficient but, unfortunately, it is inherently incomplete because there may not be any velocity profile along the selected path which is collision-free. A partial solution to this incompleteness is to make the velocity planning for several different paths [19].

Dynamic Constraints. Dealing with dynamic constraints has proved to be an intricate problem. As mentioned earlier, there are some results for exact time-optimal trajectory planning for Cartesian robots subject to bounds on their velocity and acceleration [18, 6]. Besides optimal control theory provides some exact results in the case of robots with full dynamics, i.e. Lagrangian rigid-body dynamics, moving along a given path [1, 24].

However the difficulty of the general problem and the need for practical algorithms led some authors to develop approximate methods. Their basic principle is to define a grid which is searched in order to find an optimal solution. Accordingly trajectory planning is reduced to graph search.

For instance, [23] defines a fixed grid in the workspace of a manipulator. The best path is obtained through a hierarchical search of this grid and it is then optimized in order to take into account the dynamics of the manipulator. In a similar way, [21] proposes a tessellation of the configuration space of the robot. These two approaches compute approximate time-optimal trajectories for robots with several degrees of freedom and full dynamics but their running time grows exponentially with the number of grid points. Reference [4] presents the first polynomial-time algorithm. Besides it computes a *provably good* approximate time-optimal trajectory. In other words, the closeness of the computed solution to the exact time-optimal solution is guaranteed to some tolerance ϵ. The algorithm uses the concept of *safe* trajectories, i.e. trajectories with a safety margin and is based on the construction of a grid in the state space of the robot — a mere particle in this case. References [14] and [8] extend this work to open-chain manipulators with full dynamics.

3 Statement of the Problem

The two-dimensional instance of the highway problem is stated as follows: the lanes which represent the structure of the workspace are modelled by a set of l continuous planar curves L_i, $i = 0, 2, \ldots, 2l - 2$ (the reason for this peculiar indexing will appear later). These curves are *adjacent*, i.e. arranged side by side in a parallel way with a constant distance δL. Without loss of generality straight lanes are considered. Consequently all the L_i have an equal length $p_{\max}$ and are arranged in the way depicted in Fig. 1.

A set of particles $\mathcal{B}_j$, $j = 1, \ldots, n$, represents the moving obstacles. The position of such a particle at time t is defined by the pair $(L_{\mathcal{B}_j}(t), p_{\mathcal{B}_j}(t))$ where $L_{\mathcal{B}_j}(t)$ is the index of the lane which is occupied by $\mathcal{B}_j$ and where $p_{\mathcal{B}_j}(t) \in [0, p_{\max}]$ is its curvilinear abscissa along this lane. The functions $L_{\mathcal{B}_j}(t)$ and $p_{\mathcal{B}_j}(t)$ are defined (but not necessarily everywhere) over the closed time interval $[0, t_{\max}]$ where $t_{\max}$ represents the time-horizon, i.e. the time interval over which the motions of the $\mathcal{B}_j$ are known.

The robot whose motions are to be planned is also represented by a particle $\mathcal{A}$. The normal behaviour of $\mathcal{A}$ is to follow a given lane; however $\mathcal{A}$ has also the possibility to make a lane-changing, i.e. to shift from its current lane to an adjacent one. The particular nature of these motions led us to decouple the

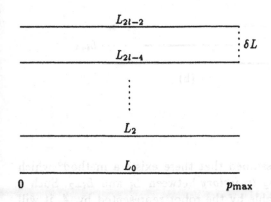

Fig. 1. the lanes L_i, $i = 0, 2, \ldots, 2l - 2$

motion along the lane — *longitudinal motion* — from the motions between the lanes — *lateral motions*.

The longitudinal motion of $\mathcal{A}$ along a lane is a one-dimensional motion which is obtained by applying an acceleration $\ddot{p}(t)$ to $\mathcal{A}$. The velocity $\dot{p}(t)$ and the position $p(t)$ of $\mathcal{A}$ along this lane are respectively defined as the first and second integral of $\ddot{p}(t)$ subject to an initial position and an initial velocity. Besides $\ddot{p}(t)$ and $\dot{p}(t)$ are bounded in the following way:

$$-\ddot{p}_{\max} \leq \ddot{p}(t) \leq \ddot{p}_{\max} \tag{1}$$

$$0 \leq \dot{p}(t) \leq \dot{p}_{\max} \tag{2}$$

Let us consider now a lateral motion, i.e. a lane-changing, as depicted in Fig. 2-a. At time t, $\mathcal{A}$ shifts smoothly from its current lane L_i to an adjacent lane L_{i+2}. The shape of the lane-changing trajectory and the time-interval Δt necessary to perform it depend on the characteristics of the robot represented by $\mathcal{A}$ and on its position and velocity at the beginning of the lane-changing.

As far as collision avoidance is concerned and assuming that the value of δL is of the order of the width of the objects represented by $\mathcal{A}$ and the $\mathcal{B}_j$s, it is reasonable that, during the lane-changing, $\mathcal{A}$ should avoid collision with the obstacles of both lanes L_i and L_{i+2}. Therefore evaluating whether the lane-changing trajectory is collision-free can be done very simply by checking out potential collision in both lanes L_i and L_{i+2} during the time interval $[t, t + \Delta t]$. This property makes it possible to model the lateral motion as a three-step process: (a) at time t, $\mathcal{A}$ instantaneously 'jumps' from L_i to a fictitious intermediate lane L_{i+1}, (b) $\mathcal{A}$ moves along L_{i+1} during $]t, t + \Delta t[$ (the obstacles of both L_i and L_{i+2} are assumed to be on L_{i+1}) and (c) at time $t + \Delta t$, $\mathcal{A}$ instantaneously 'jumps' from L_{i+1} to L_{i+2} (see Fig. 2-b). Accordingly this modelling reduces a lateral motion to a longitudinal motion along a fictitious lane.

Precisely lane-changing is dealt with in the following way: Let $p(t)$ and $\dot{p}(t)$ be respectively the position and velocity of $\mathcal{A}$ on the lane L_i at the beginning of

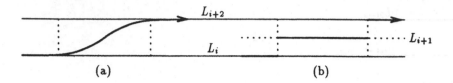

Fig. 2. lane-changing

the lane-changing ($\dot{p}(t) > 0$). It is assumed that there exists a method[3] which determines a *canonical lane-changing trajectory* between L_i and L_{i+2}. Such a canonical trajectory must be executable by the robot represented by $\mathcal{A}$, it will take a certain time Δt and it is assumed that the velocity at the end of the lane-changing is equal to the velocity at the beginning. Accordingly the fictitious acceleration $\ddot{p}(t)$ applied to $\mathcal{A}$ when it moves along the fictitious lane L_{i+1} is assumed to be null. Under these assumptions, the position and velocity of $\mathcal{A}$ at the end of the lane-changing are respectively defined as:

$$p(t + \Delta t) = p(t) + \dot{p}(t)\Delta t$$
$$\dot{p}(t + \Delta t) = \dot{p}(t)$$

In this framework, a *state* of $\mathcal{A}$ is defined as being the 3-tuple $(L, p, \dot{p})$ where $L \in \{0, 1, \ldots, 2l - 2\}$ is the index of the current lane of $\mathcal{A}$, $p \in [0, p_{max}]$ its curvilinear abscissa along this lane and $\dot{p} \in [0, \dot{p}_{max}]$ its instantaneous velocity. A *trajectory* for $\mathcal{A}$ is defined by a mapping Γ taking a time $t \in [0, t_f]$ to a state $\Gamma(t) = (L(t), p(t), \dot{p}(t))$. The *time* for the trajectory Γ is simply t_f. The two components $L(t)$ and $(p(t), \dot{p}(t))$ of this trajectory are defined by the two following maps:

1. $L: [0, t_f] \longrightarrow \{0, 1, \ldots, 2l - 2\}$ which indicates the index of the current lane of $\mathcal{A}$. If L_t is even then $\mathcal{A}$ is on the real lane L_t, otherwise it is performing a lane-changing between the lanes $L_t - 1$ and $L_t + 1$.
2. $\ddot{p}: [0, t_f] \longrightarrow [-\ddot{p}_{max}, \ddot{p}_{max}]$ which is the instantaneous acceleration applied to $\mathcal{A}$. $p(t)$ and $\dot{p}(t)$ are obtained by integration from $\ddot{p}(t)$.

Given a trajectory Γ defined by the two previous maps, the actual trajectory executed by $\mathcal{A}$ is obtained by replacing each motion along the fictitious lanes with the corresponding canonical lane-changing motion. Accordingly the time spent by $\mathcal{A}$ on a fictitious lane must be equal to Δt, i.e. the time necessary to perform the actual lane-changing motion. Besides $\ddot{p}(t)$ must be null as long as $L(t)$ is odd. A trajectory which respects these two constraints is said to be *feasible*.

As to collision avoidance, a trajectory Γ is theoretically collision-free if and only if $\forall t \in [0, t_f], p_{\mathcal{B}_j}(t) \neq p(t)$ for every $\mathcal{B}_j$ such that $L_{\mathcal{B}_j}(t) = L(t)$. However, in order to have a trajectory which be of a certain practical value, $\mathcal{A}$ should

---

[3] Such a method is described in §5 for a car-like robot.

avoid the obstacles $\mathcal{B}_j$ by a velocity-dependent safety margin $\delta(t)$. The purpose of this safety margin $\delta(t)$ is twofold: it generates a trajectory which (a) takes into account the fact that $\mathcal{A}$ and the $\mathcal{B}_j$s represent real sized objects and (b) is 'robust' in the sense that it does not skim over the obstacles (by doing so tracking errors will be allowed at execution time). Formally a trajectory Γ is said to be *safe* if and only if:

$$\forall t \in [0,\ t_f], \forall \mathcal{B}_j, j = 1, \ldots, n : L_{\mathcal{B}_j}(t) \neq L(t) \text{ or}$$
$$L_{\mathcal{B}_j}(t) = L(t) \text{ and } |p(t) - p_{\mathcal{B}_j}(t)| > \delta(t)$$

where $\delta(t) = c_0 + c_1 \mid \dot{p}(t) \mid$ with c_0 and c_1 two positive scalars.

Given an initial state $s = (L_s, p_s, \dot{p}_s)$ and a final state $g = (L_g, p_g, \dot{p}_g)$, a trajectory Γ constitutes a *solution* to the highway problem considered if and only if:

1. $L(0) = L_s$, $p(0) = p_s$, $\dot{p}(0) = \dot{p}_s$.
2. $\exists t_f \in [0, t_{\max}]$ such that $L(t_f) = L_g$, $p(t_f) = p_g$, $\dot{p}(t_f) = \dot{p}_g$.
3. $\forall t \in [0, t_f], -\ddot{p}_{\max} \leq \ddot{p}(t) \leq \ddot{p}_{\max}$ and $0 \leq \dot{p}(t) \leq \dot{p}_{\max}$.
4. $\forall t \in [0, t_f], L(t) \in \{0, 1, \ldots, 2l - 2\}$.
5. Γ is feasible.
6. Γ is safe.

The problem to be solved is to find a *time-optimal solution*, i.e. a solution Γ defined for $t \in [0, t_f]$ such that t_f should be minimal.

4 The Approach

4.1 The Basic Idea

In §3, we have defined the type of trajectory which is a solution to the highway problem. The intrinsic complexity of the problem (see §2.1) along with the time-horizon constraint led us to choose an approximate method to solve this problem. This means that, instead of trying to find out the exact time-optimal solution in the whole set of solutions, we will restrict our search to a finite subset of this set, obtaining thus an approximate time-optimal solution. The definition of this subset depends on discretizing time — a time-step τ is chosen — and selecting the acceleration of $\mathcal{A}$ from the discrete set $\{-\ddot{p}_{\max}, 0, \ddot{p}_{\max}\}$. Precisely the subset that we consider contains all the trajectories which meet the following constraints:

- $\ddot{p}$ is piecewise constant with $\ddot{p}(t) \in \{-\ddot{p}_{\max}, 0, \ddot{p}_{\max}\}$.
- $L(t)$ and $\ddot{p}(t)$ only changes their values at times $t = k\tau$ for some integer $k \geq 0$.

Such a trajectory will be called a *bang-trajectory*. It is very similar to the so-called 'bang-bang' trajectory of the control literature except that, in our case, the acceleration switches occur at regular time intervals.

Under these restrictions, the problem to be solved is to find out the time-optimal bang-trajectory. Obviously the complexity of this problem depends on the number of bang-trajectories which, in turn, is directly related to the size of τ — the smaller τ, the higher the number of bang-trajectories. On the other hand, we intuitively[4] feel that the closeness of the approximation to the exact time-optimal trajectory is also related to the size of τ — the smaller τ, the better the approximation. Thus it is possible to trade off the computation speed against the quality of the solution. However, in our case, the time-horizon constraint is the key factor which will influence the choice of τ.

In the next section, we show how to reduce the problem of finding out the time-optimal bang-trajectory to that of finding out a shortest path in a directed graph.

4.2   The State-Time Graph

A state of $\mathcal{A}$ has been defined earlier as being the 3-tuple $(L, p, \dot{p})$. A *state-time* of $\mathcal{A}$ is defined by explicitly adding the time dimension to a state of $\mathcal{A}$. Let us denote $\mathcal{TS}$ the set of all such state-times. A point in $\mathcal{TS}$ is a 4-tuple $s = (L, p, \dot{p}, t)$ or equivalently a 3-tuple $s(t) = (L(t), p(t), \dot{p}(t))$.

Motion Along a Lane. First let us consider the motion along a lane; let $s(k\tau) = (L(k\tau), p(k\tau), \dot{p}(k\tau))$ be a state-time of $\mathcal{A}$ and let $s((k+1)\tau)$ be one of the state-times that $\mathcal{A}$ can reach by a bang-trajectory of duration τ. $s((k+1)\tau)$ is obtained by applying a constant acceleration $\ddot{p} \in \{-\ddot{p}_{\max}, 0, \ddot{p}_{\max}\}$ to $\mathcal{A}$ for the duration τ. Therefore we have:

$$L((k+1)\tau) = L(k\tau)$$
$$p((k+1)\tau) = p(k\tau) + \dot{p}(k\tau)\tau + \frac{1}{2}\ddot{p}\tau^2$$
$$\dot{p}((k+1)\tau) = \dot{p}(k\tau) + \ddot{p}\tau$$

Motion Between Lanes. As mentioned earlier, $\mathcal{A}$ has also the possibility to switch to an adjacent lane. In order to do so, $\mathcal{A}$ follows a canonical lane-changing trajectory which takes a time Δt. Besides it is assumed that, in the time interval $]k\tau, (k+\mu)\tau[$, $\mathcal{A}$ moves along the fictitious lane $L(k\tau)\pm 1$ with a null acceleration. Let $\mu = \lceil \Delta t/\tau \rceil$ be the number of time-steps necessary to perform the lane-changing. The two state-times that $\mathcal{A}$ can reach at the end of the lane-changing are defined as:

$$L((k+\mu)\tau) = L(k\tau) + \sigma \text{ where } \sigma \in \{-2, 2\}$$
$$p((k+\mu)\tau) = p(k\tau) + \dot{p}(k\tau)\mu\tau$$
$$\dot{p}((k+\mu)\tau) = \dot{p}(k\tau)$$

---

[4] This intuition is confirmed in [5] where it is shown that, for a correct choice of τ, any safe trajectory can be approximated to a tolerance ϵ by a safe bang-trajectory.

The Graph. By analogy with [5], the trajectory between $s(k\tau)$ and one of the state-times that $\mathcal{A}$ can reach via a bang-trajectory of duration τ or a lane-changing trajectory of duration Δt is called a $(\sigma, \ddot{p}, \mu\tau)$-*bang* where $\sigma \in \{-2, 0, 2\}$ and $\mu \in \{1, \lceil \Delta t/\tau \rceil\}$. Let $s((k+\mu)\tau)$ be one of these state-times; it is *reachable* from $s(k\tau)$ and it is defined as:

$$L((k+\mu)\tau) = L(k\tau) + \sigma$$

$$p((k+\mu)\tau) = p(k\tau) + \dot{p}(k\tau)\mu\tau + \frac{1}{2}\ddot{p}\mu^2\tau^2$$

$$\dot{p}((k+\mu)\tau) = \dot{p}(k\tau) + \ddot{p}\mu\tau$$

Obviously a bang-trajectory is made up of a sequence of $(\sigma, \ddot{p}, \mu\tau)$-bangs. Let $s(m\tau)$, $m \geq k$, be a time-state reachable from $s(k\tau)$. Assuming that $\dot{p}(k\tau)$ is a multiple of $\ddot{p}_{\max}\tau$, we can easily show that the following relations hold for some integers α_1, α_2 and α_3:

$$L(m\tau) = L(k\tau) + \alpha_1\sigma$$

$$p(m\tau) = p(k\tau) + \alpha_2 \frac{1}{2}\ddot{p}_{\max}\tau^2$$

$$\dot{p}(m\tau) = \dot{p}(k\tau) + \alpha_3\ddot{p}_{\max}\tau$$

Thus all state-times reachable from one given state-time by a bang-trajectory lie on a regular grid embedded in TS. This grid has spacings of τ in time, of $\frac{1}{2}\ddot{p}_{\max}\tau^2$ in position and of $\ddot{p}_{\max}\tau$ in velocity and of 2 in the lane dimension.

Consequently it becomes possible to define a directed graph $\mathcal{G}$ embedded in TS. The nodes of $\mathcal{G}$ are the grid-points while the edges of $\mathcal{G}$ are $(\sigma, \ddot{p}, \mu\tau)$-bangs between pairs of these nodes. Such $(\sigma, \ddot{p}, \mu\tau)$-bangs have to respect the velocity constraint and be safe (this point is detailed in §4.3). $\mathcal{G}$ is called the *state-time graph*, it is illustrated in Fig. 3 which depicts the 'time×position×velocity' space of two adjacent lanes L_i and L_{i+2}. For the sake of clarity, we have only represented one node and its immediate neighbours on both lanes L_i and L_{i+2} (a node has at the most five neighbours). Let A be this node, the state-times reachable from A by a $(\sigma, \ddot{p}, \mu\tau)$-bang lie on the grid, they are nodes of $\mathcal{G}$. An edge between A and one of its neighbours represents the corresponding $(\sigma, \ddot{p}, \mu\tau)$-bang. A sequence of edges between two nodes defines a bang-trajectory. The time of such a bang-trajectory is defined by the time components of the two nodes considered.

Let $s = (L_s, p_s, \dot{p}_s)$ and $g = (L_g, p_g, \dot{p}_g)$ be respectively the initial and the goal state for $\mathcal{A}$. Without loss of generality it is assumed that the state-time $s^* = (L_s, p_s, \dot{p}_s, 0)$ and the set of state-times $G^* = \{(L_g, p_g, \dot{p}_g, k\tau)$ with $k = 0, \ldots, \lfloor \frac{t_{\max}}{\tau} \rfloor\}$ are grid-points. Accordingly searching for a time-optimal bang-trajectory between s and g is equivalent to searching a path in $\mathcal{G}$ between the node s^* and the node in G^* whose time component is minimal.

Note that because we only consider a compact region of TS, the number of grid-points is finite. Thus $\mathcal{G}$ is finite and the search for the time-optimal bang-trajectory can be done in a finite amount of time. The next section describes how the search is carried out.

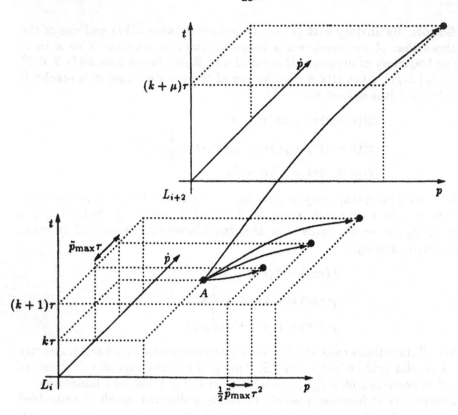

Fig. 3. the graph $\mathcal{G}$

4.3 Searching the State-Time Graph

The Algorithm. Basically we use an A^* algorithm to search $\mathcal{G}$ [17]. Beside its efficiency, such an algorithm is interesting because it generates only the parts of $\mathcal{G}$ which are relevant to the search. Starting with s^* as the current node, we 'expand' this current node, i.e. we determine all its neighbours, then we select the neighbour which is the 'best' according to a given criterion and it becomes the current node. This process is repeated until the goal is reached or until the whole graph has been explored. The time-optimal path is returned using back-pointers. The algorithm is described below:

Algorithm

Our algorithm uses a sorted list denoted by OPEN which contains the leaves of the search tree that have not yet been selected for expansion. They are sorted by increasing values of the cost function. It also keep tracks of the nodes already visited in a balanced tree structure denoted by CLOSE [26].

1. Initialize OPEN to s^* and CLOSE to $\emptyset$.

2. If OPEN is empty then return failure.
3. The head of OPEN is the new current node. Remove it from OPEN and insert it in CLOSE.
4. Compute the new nodes reachable from the current node by a valid $(\sigma, \ddot{p}, \mu\tau)$-bangs. A back-pointer is established between each of these nodes and the current node.
5. Consider each new node. If it belongs to G^*, then stop and return the time-optimal bang-trajectory using the back-pointers. Else, if the node is not in CLOSE, then insert it in OPEN according to its cost.
6. goto 2

In the next two sections, we detail two key-points of the algorithm. Namely the cost function assigned to each node and the node expansion.

The Cost Function. A^* assigns a cost $f(s)$ to every node s in G. Since we are looking for a time-optimal path, we have chosen $f(s)$ as being the estimate of the time-optimal path in G connecting s^* to G^* and passing through s. $f(s)$ is classically defined as the sum of two components $g(s)$ and $h(s)$:

- $g(s)$ is the time of the path between s^* and s, i.e. the time component of s.
- $h(s)$ is the estimate of the time-optimal path between s and an element of G^*, i.e. the amount of time it would take A to reach g from its current state with a 'bang-coast-bang' acceleration profile[5] in an obstacle-free workspace. When such an acceleration profile does not exist, $h(s)$ is set to $+\infty$.

The heuristic function $h(s)$ is trivially admissible, thus A^* is guaranteed to generate the time-optimal path whenever it exists. Besides the fact that $h(s)$ is locally consistent improves the efficiency of the algorithm [17].

The Node Expansion. The neighbours of a given node s in G are the nodes which can be reached from s by a $(\sigma, \ddot{p}, \mu\tau)$-bang. As mentioned earlier, $\sigma \in \{-2, 0, 2\}$, $\mu \in \{1, \lceil \Delta t/\tau \rceil\}$ and $\ddot{p} \in \{-\ddot{p}_{max}, 0, \ddot{p}_{max}\}$ with an extra constraint which is that $\ddot{p}$ is null and $\mu = \lceil \Delta t/\tau \rceil$ whenever $\sigma \in \{-2, 2\}$. Thus s has up to five neighbours. Let us consider the $(\sigma, \ddot{p}, \mu\tau)$-bang between $s(k\tau)$ and $s((k+\mu)\tau)$ then, $\forall t \in [k\tau, (k+\mu)\tau]$, we have:

$$p(t) = p(k\tau) + \dot{p}(k\tau)(t - k\tau) + \frac{1}{2}\ddot{p}(t - k\tau)^2 \qquad (3)$$

$$\dot{p}(t) = \dot{p}(k\tau) + \ddot{p}(t - k\tau) \qquad (4)$$

This $(\sigma, \ddot{p}, \mu\tau)$-bang is said to be *valid* if it is safe and if it does not not violate the velocity bounds (2) introduced in §3. The velocity constraint can be tested easily using (4). As to collision avoidance, we must make sure that $\forall t \in [k\tau, (k+\mu)\tau]$, $p(t)$ is no closer than $\delta(t) = c_0 + c_1 \mid \dot{p}(t) \mid$ to any obstacle of the lane $L(k\tau) + \sigma/2$. $p(t)$ is computed using (3). A practical way to check

---

[5] i.e. maximum acceleration, null acceleration and maximum deceleration.

out the safety of a given position $p(t)$ is to 'grow' of $\delta(t)$ the obstacles of the lane considered before testing whether $p(t)$ intersects the grown obstacles (see Fig. 4).

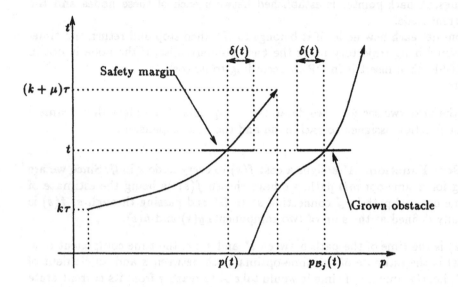

Fig. 4. safety checking

For reasons of efficiency, the safety checking within the interval $[k\tau, (k+\mu)\tau]$ is carried out by using the safety margin δ corresponding to the maximum velocity reached by $\mathcal{A}$ within this interval. In other words, we make sure that $\forall t \in [k\tau, (k+\mu)\tau]$, $p(t)$ is no closer than $\delta = c_0 + c_1 \mid \max(\dot{p}(k\tau), \dot{p}((k+\mu)\tau)) \mid$ to any obstacle of the lane considered. This conservative choice permits to grow the obstacles for a finite number of safety margins only. Namely $\delta_\alpha = c_0 + \alpha c_1(\ddot{p}_{\max}\tau)$ where $\alpha = 0, \ldots, \lfloor \frac{\dot{p}_{\max}}{\ddot{p}_{\max}\tau} \rfloor$. Therefore if we have grown the obstacles of the different values δ_α in a pre-processing phase, then checking out the safety of a single $(\sigma, \ddot{p}, \mu\tau)$-bang can be performed in time $O(n)$ where n is the total number of obstacles.

5 Implementation and Experiments

The algorithm presented above has been implemented in C on a Sun SPARC I. In the current implementation, the safety margin δ is constant, i.e. it is not speed-dependent. The next two sections detail the implementation of the safety checking procedure and the computation of a canonical lane-changing trajectory along with the time Δt necessary to perform it. Finally some experimental results are presented.

Safety Checking

A 'time×position' bitmap is used to represent the position of the grown obstacles during the time interval $[0, t_{max}]$. There is one bitmap for each real lane (see Fig. 5).

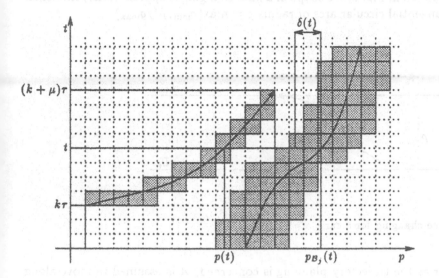

Fig. 5. the bitmap of a lane

These bitmaps enable us to efficiently check out the safety of a given $(\sigma, \ddot{p}, \mu\tau)$-bang: first the $(\sigma, \ddot{p}, \mu\tau)$-bang considered is discretized according to the resolution of the bitmaps; then each discrete point $(t, p(t))$ is tested for collision in the bitmap(s)[6] associated with the $(\sigma, \ddot{p}, \mu\tau)$-bang.

Lane-Changing

As mentioned in §3, A switches to an adjacent lane by following a canonical lane-changing trajectory. The shape of this lane-changing trajectory and the time-interval Δt necessary to perform it depend on the characteristics of the robot represented by A. In the current implementation, it is assumed that A represents a car-like robot. In order to respect the kinematic constraints of A (see [11] for more details), the gyration radius of the lane-changing trajectory must be lower-bounded by a certain value depending on (a) the minimum turning radius ρ_{min} of A (which stems from the fact that the orientation of the front wheels of a car is

---

[6] When A is performing a lane-changing between the lanes L_i and L_{i+2}, the safety of the corresponding trajectory is tested in the bitmaps of both lanes L_i and L_{i+2}.

mechanically limited), (b) the maximum centrifugal acceleration g_{max} tolerated by $\mathcal{A}$ and (c) the instantaneous velocity $\dot{p}$ of $\mathcal{A}$ ($\dot{p}$ must be strictly positive):

$$\rho \geq \max\left(\rho_{min}, \frac{\dot{p}^2}{g_{max}}\right)$$

As depicted in Fig. 6, the shape of a lane-changing is approximately modelled by two tangential circular arcs of radius $\rho = \max(\rho_{min}, \dot{p}^2/g_{max})$.

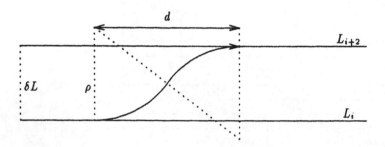

Fig. 6. lane-changing for a car

As far as the trajectory planning is concerned, $\mathcal{A}$ is assumed to move along a fictitious lane with a null acceleration. Accordingly the time Δt necessary to perform the lane-changing must be equal to $d/\dot{p}$ where d is the distance travelled along the fictitious lane:

$$d = \sqrt{\delta L(4\rho - \delta L)}$$

Experimental Results

We have tested the algorithm with up to four lanes. In these experiments, the obstacles are generated at random without caring whether they collide with each other. Besides they are assumed to keep a constant velocity on the time-horizon.

Two examples of trajectory planning involving two lanes are depicted in Figs 7 and 8. In each case, one lane is associated with two windows: a trace window showing the part of the graph which has been explored and a result window displaying the final trajectory. Any of these windows represents the 'time×position' space of the lane (the position axis is horizontal while the time axis is vertical; the frame origin is at the upper-left corner). The thick black segments represent the trails left by the moving obstacles and the little dots are points of the underlying grid. Note that the vertical spacing of the dots corresponds to the time-step τ.

In both examples, $\mathcal{A}$ starts from the first lane (lane #0), at position 0 (upper-left corner) and with a null velocity. It must reach the first lane at position p_{max} (right border) with a null velocity. $\mathcal{A}$ can overtake by using the second lane (lane

#1). In order to simulate the behaviour of a car on the roadway, we have chosen the following values for the various variables of the problem:

$$p_{max} = 500m$$
$$\delta L = 4m$$
$$\dot{p}_{max} = 72km/h$$
$$\ddot{p}_{max} = 1m/s^2$$
$$t_{max} = 100s$$

As mentioned earlier, it is the choice of τ which determines the average running time of the algorithm. For a value of τ set to 5s, we have obtained a running time ranging from less than a second to a few seconds.

6 Conclusion

In this paper, we have presented an efficient method which determines an approximate time-optimal solution to the 'highway problem', i.e. a time-optimal trajectory problem for a robot which is travelling in a structured workspace amidst moving obstacles and which is subject to constraints on its velocity and acceleration. By structured workspace, we mean that there sre lanes within which the robot is able to move. The robot has to follow a predetermined lane but it may also shift from its lane to an adjacent lane. A lane is characterized by its 'spine', i.e. a one-dimensional curve.

Our approach consists in discretizing time and selecting the acceleration applied to the robot among a discrete set. These hypotheses make it possible to define a grid in the robot's state-time space, i.e. the robot's state (or phase) space augmented of the time dimension. This grid is then searched in order to find a solution. Accordingly trajectory planning is reduced to graph search.

The choice of the time-step determines the size of the grid which, in turn, determines the average running time of the algorithm as well as how close the approximate solution is to the exact time-optimal solution. Thus it is possible to trade off the computation speed against the quality of the solution.

Acknowledgements

This work was supported by the French Ministry of Research and Technology and the European EUREKA project 'PROMETHEUS Pro-Art'.

References

1. J.E. Bobrow, S. Dubowsky, and J.S. Gibson. Time-optimal control of robotic manipulators along specified paths. *Int. Journal of Robotics Research*, 4(3):3–17, Fall 1985.

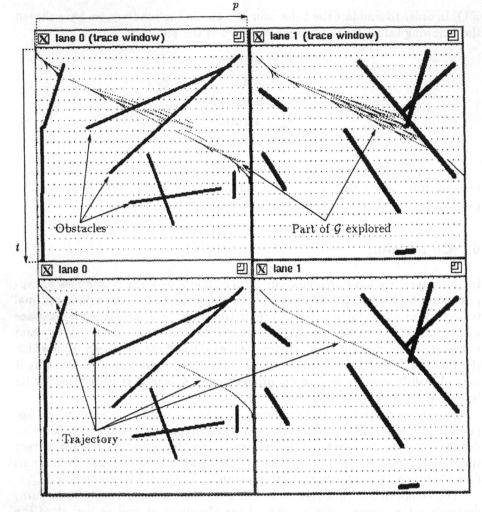

Fig. 7. an example of trajectory planning with two lanes: the solution trajectory has four lane-changings

2. R. A. Brooks. Solving the find-path problem by good representation of free space. *IEEE Trans. Systems, Man and Cybernetics*, 13(3):190–197, March/April 1983.

3. P. Caloud. Distributed motion planning and motion coordination for multiple robots. Working paper, Robotics Lab., Computer Science Dept, Stanford Univ., CA (USA), 1990.

4. J. Canny. *The complexity of robot motion planning.* MIT Press, Cambridge, MA (USA), 1988.

5. J. Canny, B. Donald, J. Reif, and P. Xavier. On the complexity of kynodynamic planning. In *Proc. of the IEEE Symp. on the Foundations of Computer Science*, pages 306–316, White Plains, NY (USA), novembre 1988.

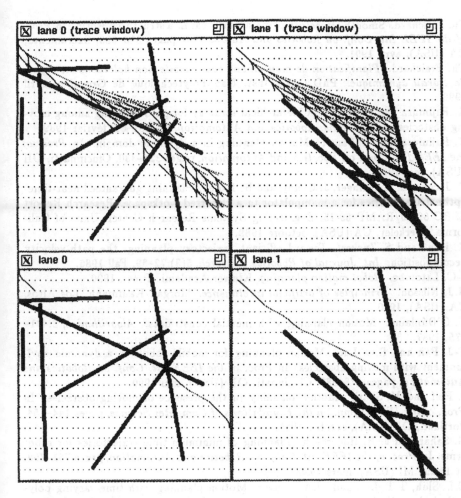

Fig. 8. an example of trajectory planning with two lanes: the solution trajectory has two lane-changings

6. J. Canny, A. Rege, and J. Reif. An exact algorithm for kinodynamic planning in the plane. In *Proc. of the ACM Symp. on Computational Geometry*, pages 271–280, Berkeley, CA (USA), 1990.

7. J. Canny and J. Reif. New lower bound techniques for robot motion planning. In *Proc. of the IEEE Symp. on the Foundations of Computer Science*, Los Angeles, CA (USA), 1987.

8. B. Donald and P. Xavier. Provably good approximation algorithms for optimal kinodynamic planning for cartesian robots and open-chain manipulators. In *Proc. of the ACM Symp. on Computational Geometry*, pages 290–300, Berkeley, CA (USA), 1990.

9. M. Erdmann and T. Lozano-Perez. On multiple moving objects. A.I. Memo 883, MIT AI Lab., Boston, MA (USA), mai 1986.

10. Th. Fraichard. Smooth trajectory planning for a car in a structured world. In *Proc. of the IEEE Int. Conf. on Robotics and Automation*, pages 318–323, Sacramento, CA (USA), avril 1991.
11. Th. Fraichard. *Planification de mouvement pour mobile non-holonome en espace de travail dynamique*. PhD thesis, Inst. Nat. Polytechnique de Grenoble, avril 1992.
12. K. Fujimura and H. Samet. A hierarchical strategy for path planning among moving obstacles. *IEEE Trans. Robotics and Automation*, 5(1):61–69, février 1989.
13. K. Fujimura and H. Samet. Motion planning in a dynamic domain. In *Proc. of the IEEE Int. Conf. on Robotics and Automation*, pages 324–330, Cincinnatti, OH (USA), mai 1990.
14. P. Jacobs, G. Heinzinger, J. Canny, and B. Paden. Planning guaranteed near-time-optimal trajectories for a manipulator in a cluterred workspace. Research Report ESRC 89-20/RAMP 89-15, Engineering Systems Research Center, Univ. of California., Berkeley, CA (USA), octobre 1989.
15. K. Kant and S. Zucker. Toward efficient trajectory planning: the path-velocity decomposition. *Int. Journal of Robotics Research*, 5(3):72–89, Fall 1986.
16. J-C. Latombe. *Robot motion planning*. Kluwer Academic Press, 1990.
17. N.J. Nilsson. *Principles of artificial intelligence*. Morgan Kaufmann, Los Altos, CA (USA), 1980.
18. C. Ó'Dúnlaing. Motion planning with inertial constraints. *Algorithmica*, 2:431–475, 1987.
19. T-J. Pan and R.C. Luo. Motion planning for mobile robots in a dynamic environment with moving obstacles. In *Proc. of the IEEE Int. Conf. on Robotics and Automation*, pages 578–583, Cincinnati, OH (USA), mai 1990.
20. J. Reif and M. Sharir. Motion planning in the presence of moving obstacles. In *Proc. of the IEEE Symp. on the Foundations of Computer Science*, pages 144–154, Portland, OR (USA), octobre 1985.
21. G. Sahar and J. H. Hollerbach. Planning of minimum-time trajectories for robot arms. In *Proc. of the IEEE Int. Conf. on Robotics and Automation*, pages 751–758, St Louis, MI (USA), mars 1985.
22. C.L. Shih, T.T. Lee, and W.A. Gruver. Motion planning with time-varying polyhedral obstacles based on graph search and mathematical programming. In *Proc. of the IEEE Int. Conf. on Robotics and Automation*, pages 331–337, Cincinnatti, OH (USA), mai 1990.
23. Z. Shiller and S. Dubowsky. Global time optimal motions of robotic manipulators in the presence of obstacles. In *Proc. of the IEEE Int. Conf. on Robotics and Automation*, pages 370–375, Philadelphia, PA (USA), avril 1988.
24. K.G. Shin and N.D. McKay. Minimum-time control of robotic manipulators with geometric path constraints. *IEEE Trans. Autom. Contr.*, 30:531–541, juin 1985.
25. G. Wilfong. Motion planning for an autonomous vehicle. In *Proc. of the IEEE Int. Conf. on Robotics and Automation*, pages 529–533, Philadelphia, PA (USA), avril 1988.
26. N. Wirth. *Algorithms and data structures*. Prentice Hall Int., Englewood Cliffs, NJ (USA), 1986.

Motion Planning for a Non-holonomic Mobile Robot on 3-dimensional Terrains

Thierry Siméon

LAAS-CNRS
7, Avenue du Colonel Roche, 31077 Toulouse Cedex, France
e-mail: nic@laas.fr

Abstract. This paper addresses the problem of autonomous navigation of mobile robots on rough terrains [1]. The contribution is a geometrical path planning algorithm that considers the geometry of the robot and its kinematic constraint in order to plan trajectories on a 3D terrain represented by polygonal patches. A first version of this planner has been implemented and simulation results that show the effectiveness of the approach are presented at the end of the paper.

1 Introduction

This paper presents a method for planning motions of a mobile robot, considering the 3-dimensional nature of the terrain, the geometry of the vehicle and its non-holonomic constraint. Potential applications of this planner include navigation of autononous mobile robots designed for outdoor environments. In particular autononous navigation is crucial for a planetary rover in rough natural terrain.

Much work on path planning has been done during the past decade for mobile robots moving in planar environments, or for manipulators (see [3] for a recent overview). Navigation on 3-dimensional terrains has been less addressed [8][7][9] and these works mainly concern the route planning problem for which the robot is considered as a point. Moreover, path planning with non-holonomic constraints is recent and has only been considered in the planar case of a polygonal robot moving amidst polygonal obstacles [1][5][6].

The paper is organized as follows: Section 2 describes the inputs of our planner. Sections 3 and 4 respectively detail the algorithms developped to solve the placement problem and to characterize the legal placements in the configuration space of the robot. The non-holonomic relation due to the kinematic constraint of the vehicle and the equations of the feasible motions are established in Sect. 5. Finally, the planner described in Sect. 6 combines these results. A first version of the planner has already been implemented. To our knowledge, this is the first planner able to solve such path-planning problems.

---

[1] This work was partially supported by C.N.E.S. (French Spatial Agency) in the framework of the French Automatic Planetary Rover project [2]

2 The World model

The terrain model defines the workspace of the robot; it is simply represented by a set of polygonal faces f_i $(i = 1, n)$ in $\mathcal{R}^3$. The vertices of these faces are given related to the fixed frame of the world R_0. The only restriction on this polyhedral model of the terrain is that for given values of both coordinates x and y, there must be an unique elevation z.

The robot (Fig. 1) considered in this work is a vehicle with two rear driving wheels and one free wheel at the front. Its body is modeled by an arbitrary polyhedron. As shown in Fig. 1, the robot frame R_{rob} is attached to the midpoint O of the line joining the centers P_1 and P_2 of the rear wheels (l equals to half the distance between these centers). The x-axis of R_{rob} is chosen to coincide with this line and the y-axis is aligned with the main axis of the robot -i.e. the line joining O to the center P_3 of the front wheel ($L = |\overrightarrow{OP_3}|$). The control parameters are the speeds V_1 and V_2 of the rear wheels. Therefore, the robot has only 2 degrees of freedom and the kinematic constraint associated with this locomotion system will be discussed in Sect. 5.

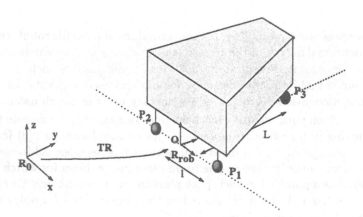

Fig. 1. Model of the vehicle

In order to simplify the contact analysis between the robot and the terrain, the contact between the wheels and the terrain are supposed to be ponctual. Moreover, these contact points are assumed to be fixed relatively to R_{rob}. This assumption implies that the radius of the wheels is small enough compared to the robot size and to the terrain irregularities. Therefore, we only consider in the sequel of the paper the case of ponctual wheels (i.e. the contact points are P_1, P_2 and P_3). This simple model of the locomotion system is used to demonstrate the approach; some extensions are discussed in the conclusion.

3 The placement problem

The placement **p** of an object in a 3-dimensional space is defined by 6 parameters $(x, y, z, \theta, \phi, \psi)$. However, the contacts between the three wheels and the terrain create relations between these parameters and a configuration vector $\mathbf{q} = (x, y, \theta)$ is sufficient to characterize this placement. For a given **q**, the placement problem consists in computing the values of the parameters z, ϕ and ψ which verify these contact relations.

The **Z-Y-X** Euler angles are used to define the orientation of R_{rob} relatively to R_0: R_{rob} is first rotated about z by an angle θ, then rotated about the new y axis by an angle ϕ and then rotated about the new x axis by an angle ψ. The situation of R_{rob} relatively to R_0 is then described by the transformation $\mathbf{TR} = Tra(x, y, z).Rot(\theta, \mathbf{z}).Rot(\phi, \mathbf{y}).Rot(\psi, \mathbf{x})$.

3.1 The contact equations

First, let us suppose that the terrain faces involved in the contact are known. Let $z = a_i x + b_i y + c_i$ be the plane equation of the face in contact with P_i $(i = 1, 3)$. These contact points expressed in R_0 are given by:

$$P_1 = \begin{pmatrix} x + l\cos\theta\cos\phi \\ y + l\sin\theta\cos\phi \\ z - l\sin\phi \end{pmatrix} \quad P_2 = \begin{pmatrix} x - l\cos\theta\cos\phi \\ y - l\sin\theta\cos\phi \\ z + l\sin\phi \end{pmatrix}$$

$$P_3 = \begin{pmatrix} x + L(\cos\theta\sin\phi\sin\psi - \sin\theta\cos\psi) \\ y + L(\sin\theta\sin\phi\sin\psi + \cos\theta\cos\psi) \\ z + L\cos\phi\sin\psi \end{pmatrix} \tag{1}$$

Considering that each contact point P_i has to verify the corresponding plane equation, the 3 following constraint relations can be easily derived:

$$2l\sin\phi + ((a_1 + a_2)c_\theta + (b_1 + b_2)s_\theta)l\cos\phi$$
$$+(a_1 - a_2)x + (b_1 - b_2)y + (c_1 - c_2) = 0 \tag{2}$$

$$2z = ((a_1 - a_2)c_\theta + (b_1 - b_2)s_\theta)l\cos\phi$$
$$+(a_1 + a_2)x + (b_1 + b_2)y + (c_1 + c_2) \tag{3}$$

$$((a_3 c_\theta + b_3 s_\theta)\sin\phi - \cos\phi)L\sin\psi$$
$$+(b_3 c_\theta - a_3 s_\theta)L\cos\psi + a_3 x + b_3 y + c_3 - z = 0 \tag{4}$$

with $c_\theta = \cos\theta$ and $s_\theta = \sin\theta$.

For a given configuration (x, y, θ) the first equation of type $A\cos\phi + B\sin\phi + C = 0$ (this type is noted $cs(\phi)$) can be analytically solved for ϕ. The elevation z is then directly obtained from the second equation for the value ϕ_{sol} of ϕ. Finally, knowing ϕ_{sol} and z_{sol}, the third equation is of type $cs(\psi)$ and can also be solved analytically for the last unknown parameter ψ. Only the roots ϕ_{sol} and ψ_{sol} inside the interval $]-\frac{\pi}{2}, \frac{\pi}{2}[$ are of interest since we only consider the placement that maintains the robot over the terrain.

3.2 Selection of potential faces

The difficulty that arises is due to the fact that the terrain faces (and therefore the equations of their support planes) cannot be known *a priori*; they depend on the values of the ϕ and ψ parameters. This section describes a two-step algorithm developped to compute the placement. **1.** A set of *candidate faces* is first computed for both rear contact points and only a limited number of couples are generated from these sets in order to calculate the angle ϕ and the elevation z. **2.** Another set of *candidate faces* is then computed for the third contact point in order to compute the angle ψ.

Figure 2 illustrates the first step of the algorithm. For a given configuration vector $\mathbf{q}$, the respective projections P_1' and P_2' onto the horizontal plane of both rear contact points P_1 and P_2 only depend on the unknown parameter ϕ. Let Q_1 (resp. Q_2) be the location of P_1' (resp. P_2') at $\phi = 0$. When the angle ϕ varies, the projected point P_i' ($i = 1, 2$) moves onto the fixed segment $[O', Q_i]$ and its position along this segment is defined by $\overrightarrow{O'P_i'} = \lambda . \overrightarrow{O'Q_i}$ with $\lambda = \cos \phi$ [2].

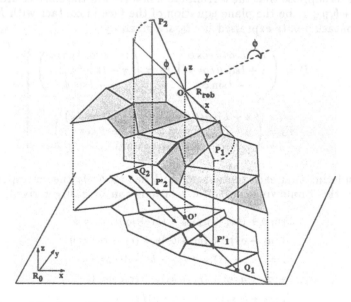

Fig. 2. Selection of the faces likely to be in contact with P_1 and P_2

Therefore, only the terrain faces whose projection intersects the segment $[O', Q_i]$ can be possibly in contact with the rear contact point P_i. These intersections with $[O', Q_i]$ occur for discrete locations of point P_i' (Fig. 3); they can be defined by a sequence of increasing values λ_i^j ($1 \leq j \leq n_i$) of λ (n_i is the number of intersections). The terrain face denoted f_i^j whose projection contains the por-

---

[2] Since $\phi \in]-\frac{\pi}{2}, \frac{\pi}{2}[, \lambda \geq 0$

tion of segment $[O', Q_i]$ corresponding to the range [3] $[\lambda_i^j, \lambda_i^{j+1}] (0 \le j \le n_i+1)$, can therefore be in contact with the contact point P_i only if $\cos \phi \in [\lambda_i^j, \lambda_i^{j+1}]$. Moreover, as P_1' and P_2' move symetrically when ϕ varies, only the couples of faces $(f_1^j, f_2^{j'})$ such that their common range $I = [\lambda_1^j, \lambda_1^{j+1}] \cap [\lambda_2^{j'}, \lambda_2^{j'+1}]$ is not empty are retained to solve for ϕ the equation (2). the solution ϕ_{sol} of the placement is obtained for the couple of faces such that the solution ϕ_{sol} of this equation verifies $\cos \phi_{sol} \in I$. The elevation z_{sol} is then computed by solving equation (3) for the same couple of faces.

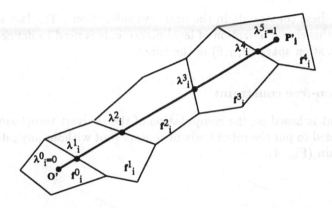

Fig. 3. Definition of f_i^j and λ_i^j

Both locations of P_1 and P_2 are now completely determined and a similar method is used to calculate the last rotation ψ needed to put P_3 in contact with the terrain. Let Q_3 be the point corresponding to P_3 for $\psi = 0$. When ψ varies, the projection of P_3 onto the plane defined by the points P_1, P_2 and Q_3 moves along the segment $[O', Q_3]$. The set of *candidate faces* for the contact with P_3 is determined by computing the faces whose projection onto this plane intersects the segment $[O', Q_3]$. These faces f_3^j $(0 \le j \le n_3+1$ in the case of n_3 intersections) are similarly associated with the range $[\lambda_3^j, \lambda_3^{j+1}]$ defining the contained portion of this segment. Finally, the last unknown angle ψ is obtained for the face f_3^j such that the solution ψ_{sol} of equation (4) verifies $\cos \psi_{sol} \in [\lambda_3^j, \lambda_3^{j+1}]$.

4 Characterization of the legal placements

Moving the robot on the terrain primarily requires a characterization of its legal placements. Contrary to classical path planners, there is no explicit description of obstacles; they are included in the terrain representation itself represented by

---

[3] λ_i^0 is null and $\lambda_i^{n_i+1}$ is equal to 1

small polygonal patches, and therefore they create irregularities that will have direct consequences on the placements of the robot. A placement p is considered to be legal if it verifies the following conditions:

- **collision-free condition:**
  This condition guarantees that the body of the robot does not collide with the terrain.
- **stability condition:**
  Only static stability is considered in this work. A placement is considered to be stable if the gravity center of the vehicle stays inside the support polygon.

We detail these constraints in the next two subsections. The last subsection explains how this characterization of legal placements is used to define obstacles in the configuration space (x, y, θ) of the robot.

4.1 Collision-free constraint

This constraint is based on the computation of the shortest translation d along the z-axis needed to put the robot body in contact (but without any intersection) with the terrain (Fig. 4).

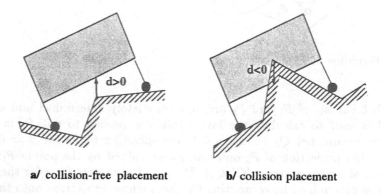

a/ collision-free placement   **b/** collision placement

Fig. 4. Definition of distance d

Let R_i $(i = 1, m)$ be the m faces of the polyhedron describing the robot body. This translation d is equal to $min_{i=1,m} d_i$ where d_i is defined by:

$$d_i = min\{z_{R_i}(x,y) - z_T(x,y) , \quad (x,y) \in proj(R_i)\}$$

$proj(X)$ is used to denote the projection of a compact set X of 3-dimensional points onto the horizontal plane and $z_{R_i}(x,y)$ (resp. $z_T(x,y)$) represents the elevation at (x,y) of the point that belongs to the robot face R_i (resp. to the terrain). Clearly, the robot body completely lies over the terrain when the following inequality is verified:

$$d > 0 \tag{5}$$

On the contrary, a negative value of d indicates that at least one face collides with the terrain. It is easy to see that the d_i values can be computed by only comparing the differences of elevations $z_{R_i}(x,y) - z_T(x,y)$ for a discrete set of points L of the x-y plane. As illustrated by Fig. 5, this set L contains:

- The intersecting points between the edges of $proj(R_i)$ and the planar subdivision obtained by projecting the terrain faces onto the horizontal plane.
- the projected vertices of the terrain that lie inside $proj(R_i)$.

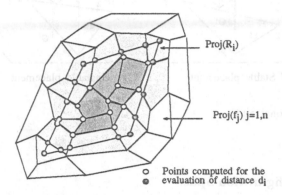

Proj(R_i)

Proj(f_j) j=1,n

○ Points computed for the
◉ evaluation of distance d_j

Fig. 5. Determination of set L

4.2 Stability constraint

The vehicle support polygon SP is defined by the projection onto the horizontal plane of the lines connecting the contact points between the wheels and the terrain. G' denotes the projection of the gravity center onto this same horizontal plane (Fig. 6). The placement is considered to be stable when the condition $G' \in SP$ is verified. This condition implies that G' has to lie inside the intersection of the half-planes defined by the lines $(P_1', P_2'), (P_2', P_3'), (P_3', P_1')$. The locations of these lines and of G' clearly depend on the placement $\mathbf{p}$ of the vehicle. However, it can be easily shown that their relative positions only depend on both orientation angles ϕ and ψ and that the stability constraint can be simply expressed as a conjonction of the three following constraints

$$\left.\begin{array}{l} \tan\psi < \frac{Y_g}{Z_g} \\ \tan\phi > \frac{-LX_g - l(L-Y_g)}{ZL}\cos\psi - \frac{l}{L}\sin\psi \\ \tan\phi < \frac{-LX_g + l(L-Y_g)}{ZL}\cos\psi + \frac{l}{L}\sin\psi \end{array}\right\} \qquad (6)$$

where X_g, Y_g and Z_g are the gravity center coordinates expressed in the robot frame.

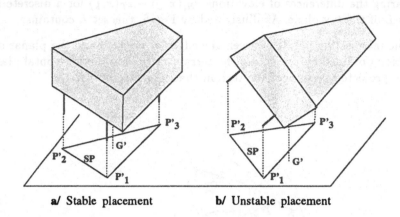

a/ Stable placement   b/ Unstable placement

Fig. 6. Stability constraint

4.3 Free Configuration Space

A **free configuration** is defined as a configuration $\mathbf{q}$ for which the associated placement $\mathbf{p}$ is legal. On the contrary, the configuration belongs to an obstacle in the configuration space if the corresponding placement is not legal.

This characterization of free placements allows to produce a discrete three dimensional configuration space by computing the placement for each possible configuration. The free regions of this configuration space correspond to regions that are crossable by the robot. However, the processing time would be important and in many cases a complete analysis of this space is not needed. The planner described in Sect. 6 allows to analyze only small regions of this space. These regions are constructed incrementally during the search for a given trajectory.

5   The kinematic constraint

The control parameters used to drive and to steer the vehicle are the speeds V_1 and V_2 of the rear wheels. Assuming a pure rolling contact between the wheels and the terrain, the velocity vectors $\mathbf{v}_1$ and $\mathbf{v}_2$ of the rear contact points P_1 and P_2 are constrained to be perpendicular to the rear axle $\mathbf{x}_{rob}$ (Fig 7). In order to maintain the contact, each $\mathbf{v}_i$ must also be perpendicular to the normal $\mathbf{n}_i$ of the corresponding contact face of the terrain. $\mathbf{v}_i$ can therefore be defined for $i = 1, 2$ by:

$$\left. \begin{array}{l} \mathbf{v}_i . \mathbf{x}_{rob} = 0 \\ |\mathbf{v}_i| = V_i \\ \mathbf{v}_i . \mathbf{n}_i = 0 \end{array} \right\} \tag{7}$$

46

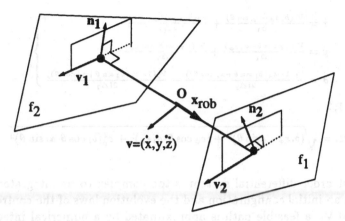

Fig. 7. Kinematic constraint

The first equation of (7) constraints the velocity vector $\mathbf{v} = (\dot{x}, \dot{y}, \dot{z})$ of point O to be also perpendicular to $\mathbf{x}_{rob}$. From the expressions of the global coordinates of $\mathbf{x}_{rob}$ [4], this constraint can be written:

$$\cos\theta\cos\phi.\dot{x} + \sin\theta\cos\phi.\dot{y} - \sin\phi.\dot{z} = 0 \qquad (8)$$

Derivating the contact equations (2) and (3) and eliminating $\dot{\phi}$ yield to an expression which relates $\dot{z}$ to the derivatives of the configuration parameters x, y and θ. Finally, substituting this expression in (8) yields to the following kinematic constraint:

$$(A_1C_2 + A_2C_1)\dot{x} + (B_1C_2 + B_2C_1)\dot{y} + lc_\phi(A_1B_2 - A_2B_1)\dot{\theta} = 0 \qquad (9)$$

$$\text{with} \quad \begin{cases} A_i = \cos\theta c_\phi - a_i s_\phi \\ B_i = \sin\theta c_\phi - b_i s_\phi \\ C_i = s_\phi(a_i\cos\theta + b_i\sin\theta) - c_\phi \end{cases} \quad (i = 1, 2)$$

The parameters $c_\phi = \cos\phi$ and $s_\phi = \sin\phi$ are themselves related to the configuration parameters by the contact equation (2). It can be shown that the kinematic constraint (9) is non holonomic (i.e. it cannot be integrated into a constraint between the configuration parameters). (9) has to be related to the classical non-holonomic constraint $\dot{y} = \dot{x}\tan\theta$ established in the case of planar motions. The consequence of any non-holonomic constraint is that an arbitrary path in the free regions of the configuration space does not necessarily correspond to a feasible trajectory for the robot. However, it has been shown [4] [1] that such a system remains controllable.

The equations of the feasible motions of the vehicle are also derived from equations (7) and from the derivation of equations (1) which relate the coordinates of both velocities $\mathbf{v}_1$ and $\mathbf{v}_2$ to the coordinates of velocity $\mathbf{v}$:

---

[4] the first column of the transformation TR

$$\left. \begin{array}{l} \dot{x} = \frac{V_1(b_1 s_\phi - c_\phi \sin\theta)}{2\Delta_1} + \frac{V_2(b_2 s_\phi - c_\phi \sin\theta)}{2\Delta_2} \\[2mm] \dot{y} = \frac{V_1(c_\phi \cos\theta - a_1 s_\phi)}{2\Delta_1} + \frac{V_2(c_\phi \cos\theta - a_2 s_\phi)}{2\Delta_2} \\[2mm] \dot{\theta} = \frac{V_1(1 - t_\phi(b_1 \sin\theta + a_1 \cos\theta))}{2l\Delta_1} - \frac{V_2(1 - t_\phi(b_2 \sin\theta + a_2 \cos\theta))}{2l\Delta_2} \end{array} \right\} \qquad (10)$$

with for $i = 1, 2$

$$\Delta_i = \sqrt{(b_i s_\phi - c_\phi \sin\theta)^2 + (c_\phi \cos\theta - a_i s_\phi)^2 + c_\phi^2 (b_i \cos\theta - a_i \sin\theta)^2}$$

This first order differential system is too complex to be integrated analytically. Given an initial configuration and the evolution laws of the control parameters V_1 and V_2, a feasible path is approximated by a numerical integration of these equations.

6 The path-planner

6.1 Overview of the planner

Given two configurations q^{init} and q^{goal}, our planner constructs a path which verifies the kinematic constraint (9), the stability and the collision-free constraints (6) (5). The planner is based on similar principles than those proposed by [1] for the planar case. It consists of exploring a discretized representation of the configuration space which is incrementally constructed during the search.

A graph is generated from the initial configuration by applying sequences of piecewise continuous inputs to the system during given time intervals. The nodes of this graph correspond to reached configurations and the arcs indicate the choice of a given control over the time interval.

The discretization of the configuration space is used to limit the size of this graph by keeping track of cells which have already been crossed by some trajectory. If a successor of the current configuration lies in a cell which has been already visited during the search, it is discarded; Otherwise it is added to the list of nodes which still need to be expanded and the corresponding cell is marked as visited. Therefore, the planner never explores regions around the trajectories which have already been computed.

6.2 Generation of the graph nodes

Let V_{max} be the maximum speed authorized for both wheels (i.e. $V_i \in [-V_{max}, V_{max}]$). The successors of the current configuration q^{cur} (reached at time t^{cur}) are generated by successively setting the value of each V_i to the elements of the set $\{-V_{max}, 0, V_{max}\}$ and by integrating the differential system (10) over a short time interval $[t^{cur}, t^{cur} + \Delta t]$. As the speeds cannot change instantaneously, the transitions between two different values of V_i^{cur} are carried out by applying over the time period Δt the linear speed:

$$V_i(t) = V_i^{cur} + \frac{V_i^{next} - V_i^{cur}}{\Delta t}.t \quad \text{for} \quad t \in [t^{cur}, t^{cur} + \Delta t]$$

Note that the coefficients of the differential system (10) depend on the equations of the terrain faces that are involved in the contact with the wheels. Due to the numerical integration, it is not possible to detect exactly when changes of the contact faces will happen. For this reason, this faces are supposed to remain the same during the time interval used for the integration. Therefore this integration time must be small enough to limit the error introduced by this approximation.

The successors for which the stability constraint or the collision-free constraint is not verified, are discarded. The forbidden regions of the configuration space are then used to truncate the graph search by cutting branches as soon as the corresponding trajectory crosses such a forbidden region.

6.3 Graph search

The search is performed by a classical A^* algorithm. To carry out the search, we must associate with each search node the actual distance covered and an underestimate of the remaining distance to the goal. The distance between two successive configurations q^{cur} and q^{next} is evaluated by the euclidian distance between the positional components of their placement vectors p^{cur} and p^{next}. The arcs that correspond to changes of the inputs values are penalized in order to limit the number of manoeuvers. In the current implementation of the system, the heuristic simply corresponds to the *flight distance* between p^{cur} and p^{goal}.

The solution of the search is a set of intermediate configurations between q^{init} and q^{goal}. The planner also returns the sequence of both V_1 and V_2 values that corresponds to the solution trajectory.

The main advantage of this planning method is that its efficiency is directly related to the complexity of the problem to be solved. On one hand, contrary to global methods there is no need for an *a priori* complete computation of the search space. On the other hand, it does not have the drawback of local methods that can lead to dead-end situations. The planner is guaranted (for a given discretisation of the search space) to find a path if it exists; it can be seen as a sort of local method which memorizes its failures by recording the trajectories generated during the search.

6.4 Experimental results

A first version of this planner has been implemented in C on a Sun SparcStation 1. An example of a trajectory computed by the system is shown in Figure 8. In this example, the terrain is modeled by 2048 faces and the solution is searched in a configuration space discretized into $64 \times 64 \times 64$ cells. The running time of the algorithm sensibly depends on how the heuristic guides the search. For this example, it ranges from a few minutes to a few tens of minutes depending on both initial and goal configurations. The trajectory of Figure 8 corresponds to one of the worst-case situations (about 20min of CPU time); Indeed, the heuristic continually attracts the search toward the cliff and an important time is spent before the planner finds out that a detour is needed.

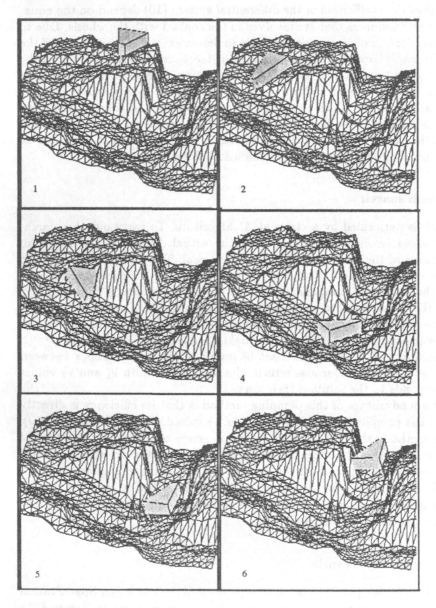

Fig. 8. Example of a path computed by the planner

7 Concluding remarks

An approach has been proposed to solve the path-planning problem for a mobile robot within a 3D environment. By characterizing the constraints specific to our problem in a three dimensional configuration space, we have proposed a path-planner based on simple techniques, similar to those developped for the planar case. In this paper, we restricted our attention on a simple model for the locomotion system of the robot. We are currently investigating the case of more realistic systems; e.g. a car-like robot with four wheels (modeled by discs) mounted on two articulated axles. These extensions lead to the introduction of several unknown variables in order to parametrize 1/ the location on each wheel of the contact point with the terrain and 2/ the degree of freedom between both axles. They significantly complicate the expression of the contact equations which cannot be solved analytically anymore. Rather than a numerical resolution of these equations, an incremental procedure which computes an approximated placement would be perhaps preferable. However, even if the placement algorithm needs to be adapted, the rest of the approach holds. The efficiency of the search could be also improved by using some global knowledge on the terrain. For example, an analysis of the terrain gradient could be used to compute uncrossable regions of the (x, y) space. Considering these regions for the evaluation of the remaining distance to the goal would help to define less underestimated heuristics and therefore, less time would be spent by the planner when important detours are needed.

References

1. J. Barraquand and J.C. Latombe. On non-holonomic mobile robots and optimal maneuvering. *Revue d'Intelligence Artificielle*, 3(2), 1989.
2. L. Boissier and G. Giralt. Autonomous Planetary Rover (VAP): The robotics concepts. In *I.A.R.P. 91 on Robotics in Space, Pisa (Italy)*,1991.
3. J.C Latombe. *Robot Motion Planning*. Kluwer Pub., 1990.
4. J.P. Laumond. Feasible trajectories for mobile robots with kinematic and environment constraints. In *International conference on autonomous systems*, Amsterdam, Netherland, 1987.
5. J.P. Laumond and T. Siméon. Motion planning for a two degrees of freedom mobile robot with towing. *LAAS/CNRS Report 89148* , Toulouse, April 1989.
6. J.P. Laumond, M. Taix, and P. Jacobs. A motion planner for car-like robots based on a mixed global/local approach. In *IEEE International Workshop On Intelligent Robots and Systems, Tsuchiura, Japan*, July 1990.
7. E. Gat, M. Slack, D.P. Miller and R.J. Firby. Path planning and execution monitoring for a Planetary rover. In *IEEE International Conference on Robotics and Automation, Cincinnati (USA)*, 1990.
8. D. Gaw and A. Meystel. Minimum time navigation of an unmanned mobile robot in a 2-1/2d world with obstacles. In *IEEE International Conference on Robotics and Automation, San Francisco (USA)*, 1986.
9. Z. Shiller and J.C. Chen. Optimal motion planning of autonomous vehicles in three dimensional terrains. In *IEEE International Conference on Robotics and Automation, Cincinnati (USA)*, 1990.

Optimal Motion Planning of a Mobile Robot on a Triangulated Terrain Model

Alain Liégeois, Christophe Moignard

L.I.R.M.M.
Université Montpellier II and C.N.R.S.
Place Eugène Bataillon
34095 Montpellier Cedex 5, France.

Abstract. This paper presents a method for generating the minimum-time path and motions of a vehicle, taking into account, the relief, the obstacles, the ground surface characteristics, and the vehicle dynamics, including engine and gears.

The terrain is modeled by a triangulation, obtained from the level curves and characteristic points and lines. Mobility coefficients, which describe the ground nature, are attributed to the geometrical elements of the triangulation in order to compute the optimal gear and speed.

The motion between two points of a mission is obtained by the use of a A^*-like search. The maximum allowable speed is computed from the vehicle dynamic model from which the driving force is obtained, taking into account the transmission : ratios and efficiencies. The terrain model provides the required resistant forces due to slope and friction.

The simulations demonstrate the system's ability for generating the orders required by a pilot : heading, gear, and speed.

1 Introduction

Most of the works so far devoted to path and motion planning of autonomous mobile robots have addressed indoor motions on homogeneous and level grounds possibly crowded with obstacles. In such cases the actuators are electric motors, the speed is low (generally assumed to be constant) so that the vehicle Dynamical equations are not called for. The planning is a purely geometric or kinematic one : searching a path -the shortest one or not- which links a starting configuration (position and heading) to a given final one while avoiding collision of the vehicle with the obstacles.

On the contrary, the scientific literature about motion planning on outdoor uneven terrains is much more scarce, despite the interesting potential applications of Advanced Robotics for surveillance, fire-fighting, agriculture, defence, planetary exploration, etc. Such a lack of results in these aeras may be due to the enormous additional theoretical and computational complexity of the related problems as compared to the two-dimensional planning :

      i) the ground surface is uneven, rugged, non-homogeneous, and more or less traversable by a vehicle depending for example upon weather conditions,

      ii) the vehicles may be powered by very complex actuators, for example by geared thermal engines,

      iii) the speeds can be high, particularly in some applications on Earth (defense, rescue, ...),

      iiii) the performance criterion is no longer (as compared to 2D problems) the minimization of the Euclidian distance. It may concern the duration of the run, or the energy consumption. Complicated constraints (hard to be modeled explicitly) -like risk, stability, etc.- must also be added.

For these reasons, new problem formulations must be seeked for. They will include

. a three-dimensional model of the ground surface,

. a model of the interactions between the vehicle and the environment (ground and atmosphere),

. planning algorithms able to cope with the above-mentioned criteria and constraints, and to find rapidly near-optimal feasible motions.

As opposed to two-dimensional problems where the kinematic constraints (non holonomy) and the vehicle dimensions have a great importance, the global planning in all-terrain conditions is most of the time allowed to discard them, that is equivalent to consider the robot as a point moving on a surface. In some cases however, when the terrain is crowded by many obstacles, the speed will be reduced and the task passed to a local planner for computing the required maneuvres.

Minimizing the duration of the run between two points defined by a higher level "mission planner" has been the main criterion chosen by the researchers interested in three-dimensional motion planning. Their contributions are summarized below.

Gaw and Meystel (GAW86) consider a terrain model based on polygonized contours ("isolines"). The vehicle is assumed to use continuously the maximum power of its thermal engine, and the optimal path is obtained by an A^* search in the graph where the nodes are the vertices of the polygonized curves. The successors of the current node are the vertices which are "visible" from it, on the same elevation curve or on one immediately below or above. The heuristic cost of A^* is computed in considering the 3D sraight line which links the current position and the goal.

Manaoui and Liégeois (MAN88, LIE90) also used discretized contours. However, the computation of the successors is here facilitated, as compared to the previous method, by a preprocessing of the data leading to a topographic description. Furthermore, a polynomial model of the engine/transmission forces and of the disturbing ones allows the computer to determine automatically the best speed and gear. The vehicle-terrain interaction is taken into account by surimposing, to the elevation map, a regular grid where each patch is labeled with a friction coefficient depending on the local ground properties at the present time. Finally, the heuristic function is improved (more informed) by computing an evaluation of the remaining time on a piecewise linear approximation of the surface (best fitting planes).

A different approach addresses the "Weighted Region Problem" (MIT87, MIT88, ROW90).The terrain is divided into homogeneous regions. In each region, say a triangle or a 2D polygon, the vehicle speed is assumed to be a constant, function of the local mobility. The time-optimal motion between two points leads one to compute a path obeying the Snell's law of light refraction (sine law). The particular case when an edge between two regions has a low cost is also treated. An approximation of the optimal path is theoretically obtained by a "continuous Dijkstra algorithm", whose computational complexity is enormous so that more practical implementations have been proposed (ROW90), but they suffer from discretization errors and still concern 2D models.

Recently, Shiller et al. (SHI90, SHI91) considered a 3D terrain and the vehicle Dynamics. The relief is modeled by bicubic patches on which a grid is superposed. The attributes of the elements of the grid correspond to the various mobility factors : drag, threats, dynamic stability, etc. The minimum-time problem is solved in two stages. The first computes an "optimal" feasible path, then the second stage computes the optimal law of motion (velocity as a function of time) on this path. The main drawback of the method is the sensitivity to the guess used to initialize continuous optimisation methods. As a limit case, no feasible path can be returned when many obstacles and steep slopes are encountered on the terrain.

Morlans and Liégeois (MOR92a, MOR92b) start from a Digital Terrain Model (D.T.M.) to build a segmentation into 3D regions whose edges are principal thalwegs (or ravines) and whose nodes are principal saddle points (or passes in the mountains). An A^*_ε algorithm computes candidate paths taking into account distance and slope, then the computed paths can be altered by finding shortcuts and for avoiding rugged zones. The method allows reactive replanning but is rather aimed towards applications where the vehicle speed is low and the environment changes are slow -which is the case in planetary exploration by robotic rovers- than for fast vehicles on Earth. Furthermore, the segmentation obtained by processing the D.T.M. is not very precise in flat parts of the surface where it must be improved by continuous approximations.

In order to obtain an admissible compromise between near-optimality and algorithmic complexity needed to solve motion planning problems which concern fast vehicles in a possibly varying 3D environment, we propose in this paper a solution based upon a triangulation of the terrain and the search for the minimum-time motion on the network defined by this triangulation. This latter tries to preserve the essential informations of a given map : contours, pits, peaks, ridges and other "characteristic" points and lines like bridges, roads and rivers, etc. (section 2). The traversable edges of the triangles are the geometric elements on which the optimal path will be looked for (at the global planning level). The triangulation ensures a reasonable amount of successors to each node (5 in average for a regularly distributed set of many nodes). The minimum time for straight-line motion along each edge is computed by using a realistic model of the engine and gears (section 3). Then an A^* algorithm is used to compute the path, the headings, the gears and the speeds (section 4). The simulations show that the optimal motion is not very sensitive, neither to parameters changes, that justifies the discretisation of the path, nor to the weighting, of the heuristic function, used for accelerating the duration of the search.

2 The Terrain Model

The terrain mapping (elevation and planimetry) is fundamental in motion planning when the vehicle Dynamics is taken into account. At every point of a path, the model must be able to supply at least the slope and the friction and drag forces, all things which are necessary for computing the speed and the duration of motion. The model must also be compatible with local and/or instantaneous hazardous factors which are likely to modify the optimal speed : for example frost, mud, smog, etc. Since we want to be sure to find, even approximately, the optimal path and motion -if they exist- between two points of the surface, a graph search method will be used, which needs a discrete terrain model. Regular grids made of rectangles or squares lead to memorise a very large number of elements, so a triangulation of the terrain is preferred here. This triangulation must include as vertices the known interesting points given by a D.T.M. or by a classical map. We are also looking for a flexible method allowing rapid data modifications and additions for refining the model in course of time, when new or more accurate information is available.

A 3D Delaunay triangulation could have been theoretically used, which has proved to be efficient in the representation of complex shapes (BOI88) and exhibits solidly established mathematical properties. It leads however to an enormous computational complexity when the surface description calls for a large number of points. Since, on Earth, most of the traversable terrains are eroded and overhangs are not likely to be encountered, we assume that every position on the map has a single value of elevation and thus a 2D triangulation will be sufficient.

2.1 Polygonation and the initial triangulation

The adequate discretization of the known contours (iso-elevation curves) of the terrain is fundamental for providing us with the data required for a good triangulation. If one does not take care, some parts of the curves may disappear and some parts of the surface may be flattened (Figure 1). In order to avoid these losses of informations, a precise polygonation of the contours is first applied then the triangulation is constrained to include not only the contour lines but also additional specific points and lines which are fundamental to describe the 3D surface as well as the characteristics of the ground in linear and surfacic elements.

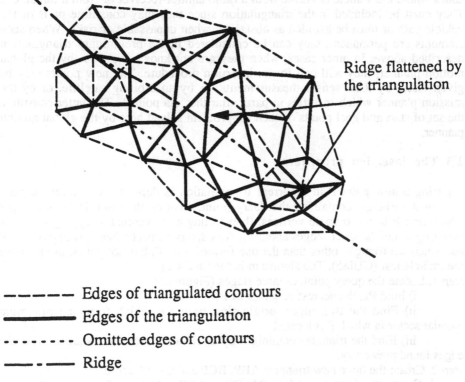

Ridge flattened by the triangulation

- – – – – – Edges of triangulated contours
- ———— Edges of the triangulation
- · · · · · · · Omitted edges of contours
- — · · — Ridge

Figure 1 : Drawbacks of an unconstrainted triangulation.

Polygonation, of the contours obtained either by map scanning or by processing a Digital Elevation Model, calls for an algorithm (MOI92a) which is similar to one among those developed for handwritten character recognition (BER79, BEL82). It uses as parameters : the length of an edge and the angle between two successive edges of the polygon under construction.

The second stage consists in a preliminary 2D (in the horizontal plane) triangulation of the set of vertices of the obtained polygons. A Watson's algorithm (SLO84) gives a Delaunay triangulation, in most cases, in an average time complexity $O(n^{3/2})$, where n is the number of points. This algorithm needs a preprocessing which consists in sorting the given points with respect to the, say, east-west direction. The result ensures that introducing the points one after one creates exactly three new edges and adds 2 to the number of triangles (three are created and one is deleted). The Delaunay quality of the triangulation is obtained by applying the "incircle test" (PRE 85, GUI85). The preprocessing allows one to create dynamically a "closed" list of triangles which no longer participate to the tests. When the convex hull of the set of points is correctly taken into account, the triangulation provides a geometrical data base consisting in n vertices, $2(n-1)-k$ triangles and $3(n-1)-k$ edges, where k is the number of points of the convex hull.

2.2 "Characteristic" points and lines

In practice, such elements are for example saddle-points (passes), pits, peaks, bridges, roads, rivers, ridges and thalwegs, borders of forests or of other types of regions, zones where the vehicle is visible from a radio emitter-receiver or from a threat, etc. They must be included in the triangulation since they may constitute parts of the vehicle path or must be avoided as obstacles when unpassable or risky. When such elements are permanent, they can be considered in the preliminary triangulation described above. In other cases, when they are not known *a priori* by the global motion planner, they will call for an insertion procedure. The new points may be given either by local sensor measurements, or by an orbiting satellite, or by the mission planner which specifies in particuliar the "via points". The latter constitute the set of start and goal points of each subproblem solved here by the global motion planner.

2.3 The insertion procedures

Inserting a new point P in a current triangulation is done in three main stages : finding the triangle containing the query point, creating the three Delaunay edges which link P to the triangle vertices, then testing and eventual swapping of edges (deleting or not "suspect" edges about P). An edge is suspect when it is opposed to P and when the triangle, other than the one formed with P, has not yet be used for the the incircle test (GUI85). The algorithm is the following :
step 1. Locate the query point, in three stages (Figure 2) :
     i) Find Pz, the nearest neighbor of P,
     ii) Find the two edges, originating from Pz, which form the minimum angular sector in which P is located
     iii) Find the triangle containig P by an iterative procedure starting from the edges found previously.
step 2. Create the three new triangles ABP, BCP and CAP (Figure 3).
step 3. Create the three new edges PA, PB and PC, which are Delaunay edges and

constitute the initial list of the neighbours of P.

step 4. Processing the point A (Figure 3)

      i) Create edge AP and add it to the neighbours of A

      ii) Sort the neighbours of A in the counterclockwise order : AD, AE, AB, AP, AC in the example.

step 5. Repeat step 4 in replacing A by B.

step 6. Repeat step 5 in replacing B by C.

step 7. Propagate the changes induced by inserting the point P : the method is similar to the one described in (GUI85) : for this purpose, test all the suspect edges about P. In Figure 4, P lies inside the circle cicumscribing the triangle CBG. Edges CB (old) and PG (new) are swapped. The algorithm must stop only after the tests of all the suspect edges, including the ones getting this label after a swapping. Furthermore, the anticlockwise listing of the edges originating from a given vertex ensures that each edge is tested only once.

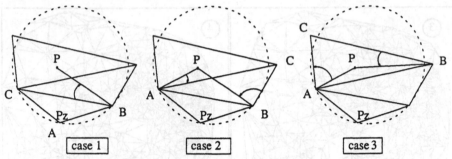

Figure 2 : Finding the triangle containing the new point P

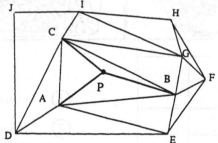

Figure 3 : Insertion of P and creation of the new triangles.

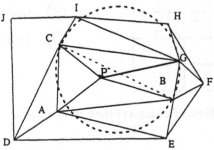

Figure 4 : Swapping a suspect edge.

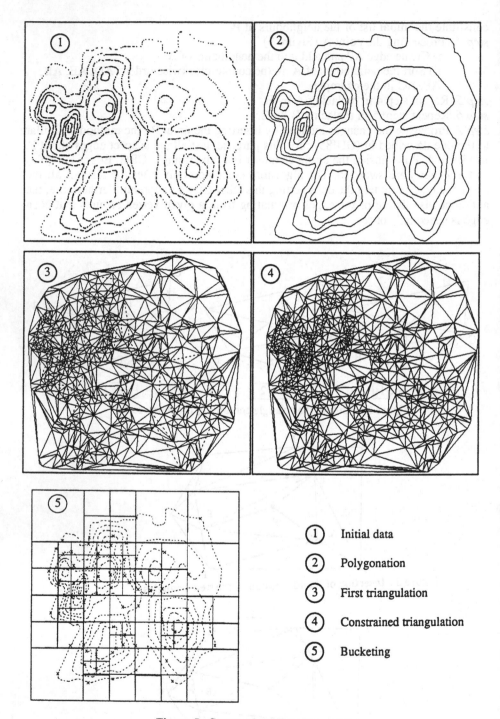

①	Initial data
②	Polygonation
③	First triangulation
④	Constrained triangulation
⑤	Bucketing

Figure 5 : Stages of the terrain modeling.

The complexity of the insertion, for most of the applications met in terrain modeling, is governed by the quality of *step 1*. For that reason, we developed a simple fast localization algorithm (MOI92a) combining a bucketing technique (OHY84, ASA88) and a gradient search in the edges graph, that gave excellent results in our applications where the number of vertices is a few thousands.

Constraining a polygonized curve to appear in the triangulation also calls for point insertion, since we choose to add the segment midpoints if necessary.
Of course, a more general and elegant solution for constructing and updating the model would have been the use of the Delaunay Tree (BOI89, DEV91), but at the expense of additional memory requirements.

2.4 Regions

The regions must be in general constructed interactively by the human specialists who designate sets of triangles and give them some attributes. Two kinds of maps are added by this way to the surface model. The first map describes the ground nature (the attribute in each region is a friction coefficient which will be defined in the next section). The second map is composed of regions where the attribute is a risk coefficient which will penalise some path segments and make other preferable.

Figure 5 illustrates the stages in the model construction. More details are found in (MOI92a).

3 Vehicle Dynamic Model and Speed Computation

The minimum-time motion computation requires the knowledge of the optimal gear-ratio and speed on each segment of a trajectory approximated by a polygonal line formed by consecutive edges of the triangulation. More precisely, the vehicle is assumed to follow line segments parallel to the corresponding edges, located at an infinitely small distance from them on the triangles where the vehicle pitch axis is the most horizontal, in order to prevent slippage or tip-over (SHI90).

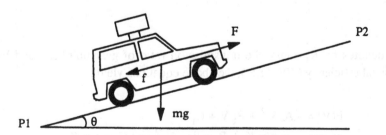

Figure 6: Forces acting on the vehicle in the forward motion.

The dynamics can be expressed as :

$$m \, dV/dt = F(V) - f(V) - m \, g \, (\sin \theta + K_f \cos \theta) \qquad (1)$$

where m is the vehicle mass, g the acceleration of gravity, K_f the friction coefficient on the path segment, V the vehicle speed, t the time and θ denotes the angle of the slope with the horizontal plane (Figure 6). F(V) is the active force and f(V) includes other types of friction, for example the aerodynamic drag, which can be modelled as polynomials : $f(V) = aV^2+bV+c$. By neglecting the transients, dV/dt is set to zero so that V is the positive maximal solution, if exists, of a polynomial equation as long as F(V) is polynomial. It is For this purpose that the torque and power versus engine speed characteristics are considered. For most of the atmospheric engines, it is found that the torque characteristic $C(\omega)$ is fairly well approximated by a parabola (Figure 7), which corresponds in turn to a third-order polynomial for the power $P(\omega)$:

$$C(\omega) = - \alpha \, \omega^2 + \beta \, \omega + \gamma \qquad (2)$$

$$P(\omega) = \omega \, C(\omega). \qquad (3)$$

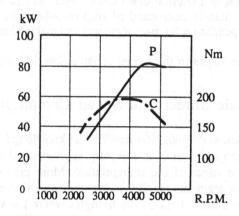

Figure 7 : **Engine Power and Torque**

If one denotes by $K_i = \omega/V$ the transmission ratio of gear number i and by r_i the mechanical efficiency ($FV = r_i P$), the above equations yield :

$$F(V) = - A_i \, V^2 + B_i \, V + C_i \qquad (4)$$
where $A_i = \alpha \, r_i \, K_i^3$, $B_i = \beta \, r_i \, K_i^2$ and $C_i = \gamma \, r_i \, K_i$.

At this stage, it can be noted that the coefficients α, β and γ can be computed either by identification of experimental curves or by the knowledge of the engine maximum torque and power, and at least of one of the corresponding speeds.
Then, the friction coefficients are subtracted :

$$A_i = A_i + a, \quad B_i = B_i - b, \quad C_i = C_i - c.$$

In order to compute the maximum speed along a climbing path, the algorithm works as follows (LIE92):

1. select the high gear : $i = i_{max}$,

2. (loop) compute the best solution of $-A_i V^2 + B_i V + C_i - mg(\sin\theta + Kf\cos\theta) = 0$.

     If it is real then return i and the best speed consistent with the maximum allowed RPM else $i = i-1$.

     If i=0 then return "infinite cost", else continue.

For descending a slope, the same algorithm can be used (replacing $\sin\theta$ by $| \sin(\theta) |$) else one can decide to select the maximum allowable speed, which is more risky. In the former case, the computed speed is :

$$V_{comp} = (B_i + (B_i^2 + 4A_i(C_i - mg(|\sin \theta| + K_f \cos \theta)))^{1/2})/2A_i \qquad (5)$$

and the optimal speed is

$$V = V_{max} \text{ if } V_{comp} > V_{max}, \text{ else } V = V_{comp} \qquad (6)$$

where

$$V_{max} = \omega_{max}/K_i, \text{ with } \omega_{max} = (\beta + (\beta^2 + 3\,\alpha\gamma)^{1/2})/3\alpha \qquad (7)$$

Finally, the time is computed by $T = D/V$, where D is the segment length. It is demonstrated in (MOI92b) that the computation is also valid when parabolic transitions between the consecutive segments are considered.

4 Motion Planner Implementation and Examples

The motion planner is responsible for computing the time-optimal path from a starting point to a goal point (both added to the triangulated terrain model) in following some edges of the triangulation. It provides also the corresponding gear ratios and speeds. For that purpose, an A^* algorithm is called for, which uses classically $f = g + h$ as the cost function, where g is the computed time from the starting point to the current node (vertex) and h is some evaluation of the minimum time required for reaching the goal. In the general case of the application considered in this paper, the problem may not be symmetric, i.e. two different costs can be associated with the same part of a path, depending upon the direction of motion, since different strategies for descending slopes can be used. At each step of the algorithm which builds the search graph, the cost functions g and h of all the successors of the current node are computed. The triangulation used for modelling the terrain ensures that this is done in constant time, since it is known (PRE85) that the mean number of edges originating from a vertex is equal to six. This result, associated with the fast computation of adding eventually on line new vertices, makes the proposed method more able to cope with uncertainties (reflexive planning) than the others (GAW86, MAN88) where the number of successors ("visible nodes") and the computing time cannot be easily estimated.

Another feature lies upon the choice of the heuristic evaluation function h. It has been proved (NIL84) that if $h < h^*$ where h^* is the minimum cost to the goal, the A^*

algorithm is admissible, i.e. it finds the optimum, if it exists. In the case considered here, setting $h = h_0$ provides admissibility, where $h_0 = D_0/V$, D_0 being the straight-line distance between the current point and he goal, and V chosen by one of the following methods:

1) $V=V_{max}$,

2) V is computed by equation (5), assuming an hypothetical straight-line motion in the three-dimensional space and taking the minimum value of K_f.

The time given by 1) is the inferior limit, that provides a h_0 much more less than h^* and leads A* to construct a leafy search graph. For that reason, solution 2) is preferred. Moreover, since getting rapidly a path is more important in practice than ensuring exact optimality (taking into account the previous approximations concerning the terrain discretisation and the dynamic transients), the modified cost function evaluation $f = g + kh_0$ has been used, where $k>1$.

As it is known, the corresponding algorithm works as pure A^* when $k = 1$ and follows the gradient of h_0 when k is large. Increasing k thus reduces the number of nodes developed during the search, and consequently, the computation time. A large number of experiments confirmed this behavior and a "good" choice of k was looked for. Figure 8 gives an example of the CPU time as a function of k. It has been found that $k = 1.2$ almost provides the optimal result and that $k=1.5$ induces only a 3% error, which is tolerable while the computing time is reduced by a factor of ten.

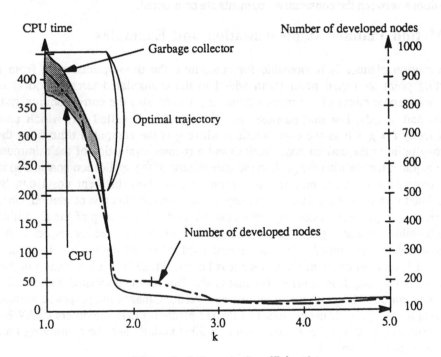

Figure 8 : Influence of coefficient k .

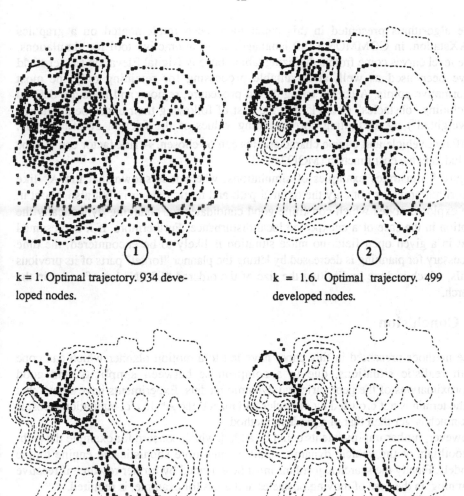

k = 1. Optimal trajectory. 934 developed nodes.

k = 1.6. Optimal trajectory. 499 developed nodes.

k = 2.2. Near–Optimal trajectory (error = 5%). 200 developed nodes.

k = 3.6. Non Optimal trajectory. 122 developed nodes.

Figure 9 : The explored nodes function of k.

63

The algorithms presented in this paper have been implemented on a graphics VAXstation, in COMMON LISP language, as an interactive tool for simulations. The level curves come from a digitizing tablet, but any Digital Elevation Model could have been used as well, after suitable processing for retaining only the most informative points. After the triangulation procedure, regions and lines are defined (including "obstacles" in the large), the cost of following their edges being altered ("weighted") by attributing corresponding values of the friction coefficient K_f.

Different vehicle and engine characteristics can be chosen. A risk factor can also be included in the cost function (MOI92a).

Figure 9 illustrates examples of simulations, where it is shown that the A^*-like algorithm provides not only the optimal path and motion along it, but also, about it, the explored nodes which constitute good candidates for changing dynamically the motion in the case of a unexpected local disturbance. Furthermore, if it is presumed that in a given open field, no maze situation is likely to be encountered, the time necessary for planning is decreased by letting the planner "forget" parts of its previous trials, which is done by limiting the size of the ordered "OPEN" list of nodes in the search.

5 Conclusion

The methods presented in this paper have led to a motion planner which can cope with realistic situations, due to a compromise between simplicity and good approximations of the geometric and dynamic models. Experiments have shown that if the terrain is not crowded by obstacles and mazes, the A^*-like computational cost is not much larger than that of a gradient method.

However, further improvements could be added, which concern for example the path-smoothing, or the vehicle dynamic stability, and more sophisticated engine torque models. Finally the search algorithms must be extended by further research if reactive planning, in the case of moving obstacles and goals for example, is required.

Acknowledgements

Parts of this work have been supported by D.R.E.T.-D.G.A. under contracts No 86/117 and 89/445.

References

(ASA88)   T. Asano: Practical use of Bucketing Techniques in Computational Geometry. Research Memorandum, Dep. of Mathematical Engineering and Instrumentation Physics, University of Tokyo, 1988.

(BEL82)   A. Belaïd, G. Masini: Segmentation de tracés sur tablette graphique en vue de leur reconnaissance. T.S.I.,Vol. 1, No 2, 1982, pp. 155-168.

(BER79)   M. Berthod, P. Jancène: Le prétraitement des tracés manuscrits sur une tablette graphique. Proc. Congrès AFCET-INRIA Reconnaissance de Formes et Intelligence Artificielle, Toulouse, Sept. 1979, pp. 195-205.

(BOI88)   J.D. Boissonnat. Shape reconstruction from planar cross sections. Computer Vision, Graphics and Image Processing, 44, 1988, pp. 1-29.

(BOI89)   O. Boissonnat, M. Devillers-Teilland: On the Randomized Construction of the Delaunay Tree. INRIA Research Report No 1140, Dec. 1989.

(DEF85)   L. De Fioriani, B. Falcidieno, C. Pienovi: Delaunay-based Representation of surfaces Defined over Arbitrarily Shaped Domains. Computer Vision, Graphics and Image Processing, 32, 1985, pp. 127-140.

(DEV90)   O. Devillers, S. Meiser, M. Teilland: Fully Dynamic Delaunay Triangulation in Logarithmic Expected Time per Operation: INRIA Research Report No 1349, Dec. 1990.

(GAW86)   D. Gaw, A. Meystel: Minimum-Time Navigation of an Unmanned Mobile Robot in a 2-1/2D World with Obstacles. Proc. 1986 IEEE Int. Conf. on Robotics and Automation, vol. 3, pp. 1670-1677.

(GUI85)   L. Guibas, J.Stolfi: Primitives for the Manipulation of General Subdivisions and the Computation of Voronoï Diagrams. ACM Trans. on Graphics, vol. 4, No 2, April 1985, pp. 74-123.

(KIR83)   D. Kirkpatrick: Optimal search in Planar Subdivisions. SIAM J. Comput., vol. 12, No 1, February 1983, pp. 28-35.

(LIE89)   A. Liégeois: Emulation d'algorithmes optimaux de mouvements de mécanismes complexes en espace encombré. Rapport de synthèse finale, Convention DRET-LAMM No 86/117, June 1989.

(LIE92)   A. Liégeois, C. Moignard: Minimum-Time Motion Planner for Mobile Robots on Uneven Terrains. In : S.G. Tsafestas (ed.), Robotic Systems, Advanced Techniques and Applications, 1992 Kluwer Academic Publishers, pp. 271-278.

(MAN88)   F. O. Manaoui: Etude et simulation d'algorithmes de navigation pour robots mobiles autonomes sur terrain inégal. Thèse de Doctorat, Université Montpellier II, France, December 1988.

(MIT87)   J.S.B. Mitchell, C.H. Papadimitriou: The Weighted Region Problem (extended abstract). Proc. 1987 ACM Conf. on Computational Geometry, pp. 30-38.

(MIT88)   J.S.B. Mitchell: An algorithmic approach to some problems in terrain navigation. Artificial Intelligence, 37, 1988, pp. 171-201.

(MOI92a)   C. Moignard: Planification et génération de mouvements en terrain inégal pour véhicule autonome à moteur thermique. Thèse de Doctorat, Université Montpellier II, France, July 1992.

(MOI92b)   C. Moignard, A. Liégeois: Planification de mouvements optimaux de robots mobiles sur terrain inégal. APII, Vol.26, No 4, 1992, pp. 39-58.

(MOR92a)   R. Morlans: Génération de trajectoires d'un robot d'exploration planétaire utilisant un Modèle Numérique de Terrain issu d'images prises par satellite. Thèse de Doctorat, Université Montpellier II, France, June 1992.

(MOR92b)   R. Morlans, A. Liégeois: A DTM-Based Path Planning Method for

Planetary Rovers. Proc. Int. Symp. on Missions, Technologies and Design of Planetary Mobile Vehicles, Toulouse, France, June 28-30 1992. Paper 4-10. CEPADUES Ed. 1992.

(NIL84)   N. J. Nilsson: Principles of Artificial Intelligence (Springer-Verlag, 1984).

(OHY84)   T. Ohya, M. Iri, K. Murota: A fast Voronoi-Diagram Algorithm with Quaternary Tree Bucketting. Information Processing Letters 18, 1984, pp. 227-231.

(PET87)   G. Petrie, T.J.M. Kennie: Terrain modelling in Surveying and Civil Engineering. Computer Aided Design, vol. 19, No 4, May 1987, pp. 171-187.

(PRE85)   F.P. Preparata, M.I. Shamos: Computational Geometry-An Introduction (Springer-Verlag, 1985).

(ROW90)   N.C. Rowe, R.F. Richbourg: An Efficient Snell's Law Method for Optimal-Path Planning across Multiple Two-Dimensional Irregular, Homogeneous-Cost Regions. The Int. J. of Robotics Research, vol. 9, No 6, December 1990, pp. 48-66.

(SAM82)   H. Samet: Neighbour finding techniques for images represented by quad-trees. Computer Graphics and Image Processing, No 18, 1982, pp. 37-42.

(SHI90)   Z. Shiller, J.C. Chen: Optimal Motion Planning of Autonomous Vehicles in Three Dimensional Terrains. Proc. 1990 IEEE Int. Conf. on Robotics and Automation, vol. 1, pp. 198-205.

(SHI91)   Z. Shiller, Y.R. Gwo: Optimal Motion Planning of Autonomous Vehicles. IEEE Trans. on Robotics and Automation, Vol. 7, No 2, April 1991, pp. 241-249.

(SLO84)   S.W. Sloan, G.T. Houlsby: An implementation of Watson's algorithm for computing 2-dimensional Delaunay triangulations. Adv. Eng. Software, 1984, vol. 6, No 4, pp. 192-197.

Motion Planning and Control under Uncertainty Constraints

Landmark-Based Robot Motion Planning

Anthony Lazanas and Jean-Claude Latombe

Robotics Laboratory, Department of Computer Science
Stanford University, Stanford, CA 94305, USA

Abstract. This paper considers a reduced version of the general motion planning problem with uncertainty, and an implemented complete polynomial algorithm solving it. This algorithm computes a guaranteed plan by backchaining nondirectional preimages of the goal until one fully contains the set of possible initial positions of the robot. It assumes that "landmarks" are scattered across the workspace. Robot control and sensing are perfect within the fields of influence of these landmarks, while control is imperfect and sensing null outside these fields. We propose extensions of the planning algorithm that eliminate the need for several of these assumptions.

1 Introduction

Robots must deal with errors in control and sensing. Since the general motion planning problem in the presence of uncertainty seems intrinsically hard [3], a promising line of research is to identify a restricted, but still useful, subclass of problems that can be solved in polynomial time.

We consider here a class of planning problems in the context of the navigation of a mobile robot. We assume that landmarks are scattered across the robot's two-dimensional workspace. Each landmark is a feature, or a combination of features, that the robot can sense and identify if it is located in some appropriate subset of the workspace [12]. A landmark may be a pre-existing feature (e.g. the corner made by two walls), or an artificial one specifically provided to help robot navigation (e.g. a radio beacon or a magnetic device buried in the ground). This subset is the field of influence of the landmark. We assume that robot control and sensing are perfect in ... field of influence of the landmark, and that control is imperfect and sensing is null outside all such fields. Given an initial region where the robot is known to be and a goal region where we would like the robot to get, the problem is to plan motion commands whose execution guarantees that the robot will move into the goal and stop there. The motion commands should also prevent the robot from colliding with stationary obstacles placed in the workspace. This problem is a simplification of a real mobile robot navigation problem, but it captures its most important aspects.

We describe an implemented planning method that backchains nondirectional preimages (weaker preconditions) of the goal until one preimage encloses the

Landmark-Based
Robot Motion Planning

Anthony Lazanas and Jean-Claude Latombe

Robotics Laboratory, Department of Computer Science
Stanford University, Stanford, CA 94305, USA

Abstract. This paper considers a reduced version of the general motion planning problem with uncertainty and an implemented complete polynomial algorithm solving it. This algorithm computes a guaranteed plan by backchaining nondirectional preimages of the goal until one fully contains the set of possible initial positions of the robot. It assumes that "landmarks" are scattered across the workspace. Robot control and sensing are perfect within the fields of influence of these landmarks, while control is imperfect and sensing null outside these fields. We propose extensions of the planning algorithm that eliminate the need for several of these assumptions.

1 Introduction

Robots must deal with errors in control and sensing. Since the general motion planning problem in the presence of uncertainty seems intrinsically hard [3], a promising line of research is to identify a restricted, but still useful subclass of problems that can be solved in polynomial time.

We consider here a class of planning problems in the context of the navigation of a mobile robot. We assume that *landmarks* are scattered across the robot's two-dimensional workspace. Each landmark is a feature, or a combination of features, that the robot can sense and identify if it is located in some appropriate subset of the workspace [13]. A landmark may be a pre-existing feature (e.g., the corner made by two walls) or an artificial one specifically provided to help robot navigation (e.g., a radio beacon or a magnetic device buried in the ground). This subset is the "field of influence" of the landmark. We assume that robot control and sensing are perfect in the fields of influence of the landmarks, and that control is imperfect and sensing is null outside all such fields. Given an initial region where the robot is known to be, and a goal region where we would like the robot to go, the problem is to plan motion commands whose execution guarantees that the robot will move into the goal and stop there. The motion commands should also prevent the robot from colliding with stationary obstacles placed in the workspace. This problem is a simplification of a real mobile-robot navigation problem, but it captures its most important aspects.

We describe an implemented planning method that *backchains nondirectional preimages* (weakest preconditions) of the goal, until one preimage encloses the initial region [14]. Each nondirectional preimage is computed as a set of directional preimages for critical directions of motion. At every iteration, the intersection of the nondirectional preimage with the fields of influence of the landmarks define the intermediate goal from which to backchain. The algorithm takes polynomial time in the total number of landmarks and obstacles. It is complete with respect to the problems it attacks, that is, it produces a guaranteed plan, whenever one such plan exists, and returns failure, otherwise. Interestingly, once a motion plan has been generated, the assumption that control and sensing are perfect in landmark areas can be relaxed to some extent without losing plan guaranteedness. Actually, in Section 9 we will discuss several extensions of the planner that eliminate the most restricting assumptions of our problem.

Previous methods to compute motion plans under uncertainty include the non-implemented algorithms of [6, 4], which take exponential time in the size of the input problem, and the implemented algorithm described of [10], which is both exponential and incomplete. The polynomiality and completeness of our algorithm derive from the combination of the two notions of a landmark and a nondirectional preimage. Since landmarks require the workspace to include appropriate features and the robot to be equipped with sensors able to detect them, this combination is an illustration of the interplay between engineering and algorithmic complexity in robotics. As software becomes more critical in modern robots, the importance of this interplay will increase. Robots will have to be designed with the tractability of their programming in mind.

2   Background

Motion planning with uncertainty is a critical problem in robotics (e.g., in mechanical assembly and mobile robot navigation tasks). It is also a notoriously hard problem [3, 4, 6, 16]. A classical approach to this problem is preimage backchaining [14]. A motion command is described by a *commanded direction of motion d* and a *termination condition* **TC**. When such a command is executed in free space, the robot moves along a direction contained at each instant in a cone of half-angle θ about d, the *directional uncertainty* (maximal error) of the robot. The motion stops when **TC** becomes true. Typically **TC** depends on sensory data, which are not accurate.

The *preimage* of a goal region for a given motion command $M = (d, \textbf{TC})$ is the set of all points in the robot's configuration space such that if the robot starts executing the command from one of these points, it is guaranteed to reach the goal and stop in it. Preimage backchaining consists of constructing a sequence of motion commands M_i, $i = 1, \ldots, n$, such that, if $\mathcal{P}_n$ is the preimage of the goal for M_n, $\mathcal{P}_{n-1}$ the preimage of $\mathcal{P}_n$ for M_{n-1}, and so on, then $\mathcal{P}_1$ contains the initial region where the robot is known to be prior to executing the plan.

One source of difficulty is the interaction between goal reachability and goal recognizability. The robot must both reach the goal (despite directional uncertainty) and stop in the goal (despite sensing uncertainty). This interaction seems to make everything depend on everything else. In particular, goal recognition, hence the termination condition of a command, often depends on the region from where the command is executed. This region, which is precisely the preimage of the goal for that command, also depends on the termination condition. This recursive dependence was noted in [14]. Despite this difficulty, Canny [4] described a complete planner with very few assumptions in it. But this planner takes double exponential time in the number of steps of the generated plan and this number can grow exponentially in the complexity of the environment.

At the expense of completeness, Erdmann [8] suggested that goal reachability and recognizability be treated separately by identifying a subset of the goal, called a *kernel*, such that when this subset is attained, goal achievement can be recognized independently of the way it has been achieved. He defined the backprojection of a region $\mathcal{R}$ for a command $\mathbf{M}$ as the set of points such that if the robot starts executing $\mathbf{M}$ from any one of these points it is guaranteed to reach $\mathcal{R}$. He proposed an $O(n \log n)$ algorithm to compute backprojections in the plane when the obstacles are polygons bounded by n edges. An implemented planner based on this approach is described in [10].

Once the kernel of a goal has been identified, a remaining problem is the selection of the commanded direction of motion to attain this kernel, since the backprojection of the kernel varies when the direction changes. In the plane, Donald [6] showed that in order to describe the set of backprojections of a region for all possible directions of motion, the *nondirectional backprojection*, it is sufficient to compute backprojections for a finite number of directions obtained by identifying critical directions where the topology of the backprojection's boundary changes. He proposed an $O(n^4 \log n)$ algorithm, with n being the number of obstacle edges, to compute nondirectional backprojections and embedded it into a polynomial one-step planner. Briggs [1] reduced the complexity of this algorithm to $O(n^2 \log n)$.

Even when nondirectional backprojections are used, a remaining difficulty to construct a multi-step planner is backchaining. Donald proposed an algorithm that is exponential in the number of steps [6]. The difficulty comes from the fact that backchaining introduces a twofold variation: when the commanded direction of motion varies, both the backprojection of the current kernel and the kernel of this backprojection (which will be used at the next backchaining iteration) vary. It is still an open problem to efficiently compute the nondirectional preimage of a nondirectional backprojection.

In this paper we directly address the nondirectional preimage backchaining problem in multiple-step planning. We deal with the difficulty mentioned above by introducing landmarks forming "islands of perfection". This notion is similar to those of "atomic region" [2], "signature neighborhoods" [15], and "perceptual equivalent classes" [5, 7]. As we will show in the rest of this paper, landmarks allow us to reduce nondirectional preimage backchaining to iteratively computing the nondirectional backprojection of a growing set of landmark regions.

3 Problem Statement

The robot is a point moving in a plane, the *workspace*, containing stationary forbidden circular regions, the *obstacle disks*. The robot can move in either one of two control modes, the *perfect* and the *imperfect* modes.

The perfect control mode can only be used in some stationary circular regions, the *landmark disks*, modeling the fields of influence of the landmarks. These disks have null intersection with the obstacle disks (both types of disks are closed subsets). When the robot is in a landmark disk, it knows its position exactly and it has perfect control over its motions. Some landmark disks may intersect each other, creating larger areas, called *landmark areas*, through which the robot can move in the perfect control mode. A motion command in the perfect control mode is called a *P-command*.

A motion command in the imperfect control mode, called an *I-command*, is described by a pair $(d, \mathcal{L})$, where $d \in S^1$ is a direction in the plane (the *commanded direction*) and $\mathcal{L}$ is a set of landmark disks (the *termination set*). This command can be executed from anywhere in the plane outside the obstacle disks. The robot follows a path whose tangent at any point makes an angle with the direction d that is no greater than some prespecified angle θ (the *directional uncertainty*). The robot stops as soon as it enters a landmark disk in $\mathcal{L}$.

The robot has no sense of time, which means that the modulus of its velocity is irrelevant to the planning problem.

At planning time, the initial position of the robot is known to be anywhere in a specified *initial region* $\mathcal{I}$ that consists of one or several disks. Each initial-region disk may be disjoint from the landmark areas, or it may overlap some of them, or it may be entirely contained in one of them. The robot must move into a given *goal region* $\mathcal{G}_0$, which is any subset of the workspace whose intersection with the landmark disks is easily computable. The problem is to generate a *guaranteed motion plan*, i.e., an algorithm made up of I- and P-commands whose execution guarantees that the robot will be in $\mathcal{G}_0$ when the execution of the plan terminates. The robot is not allowed to collide with any of the obstacle disks.

In the sequel we assume that there are ℓ disks (landmarks and obstacles) in total. We also assume that the initial region consists of $O(1)$ disks and that the goal region is simple enough so that computing its intersection with k disks takes $O(k)$ time.

4 Example

Fig. 1 illustrates the previous statement with an example run using the implemented planner. The workspace contains 23 landmark disks (shown white or grey) forming 19 landmark areas, and 25 obstacle disks (shown black). The directional uncertainty θ is set to 0.09 radian. The initial and goal regions are two small disks designated by $\mathcal{I}$ and $\mathcal{G}_0$, respectively. The white landmark disks are those with which the planner has associated I-commands. The arrow attached to a white disk is the commanded direction of motion of an I-command planned

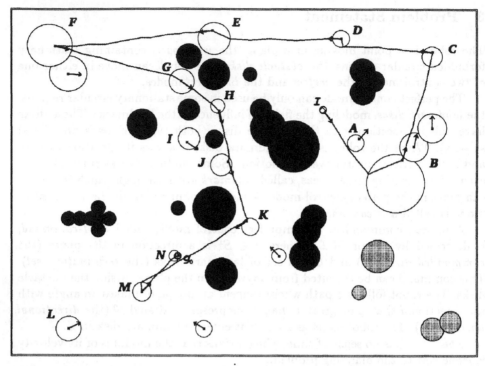

Fig. 1. Example of a planning problem

to attain another set of disks. There is at least one arrow per white landmark area not intersecting the goal.

Execution begins with performing the I-command attached to $\mathcal{I}$. When the robot reaches a landmark disk in the termination set of this command, it is guaranteed that a P-command is attached to this disk to attain a point in the current landmark area that is either a goal point (if the goal region intersects this landmark area) or such that an I-command is associated with it (the arrows shown in the figure are drawn from such points). In the first case, plan execution terminates when the goal point is attained. In the second case, the I-command is executed, and so on.

The figure also shows the path produced by a sample execution of the plan. This path first takes the robot from the initial region to the landmark area designated by B. From there, it successively attains and traverses the landmark areas marked C, D, E, F, G, H, J, K, M, and N. The P-command associated with N takes the robot to $\mathcal{G}_0$ where it stops. Due to control errors, other executions of the same plan could produce other paths traversing different landmark disks.

5 Directional Preimage

Consider a goal region $\mathcal{G}$. We define the *kernel* of $\mathcal{G}$ as the largest set of landmark disks such that, if the robot is in one of them, it can attain the goal by executing a single P-command.[1] Thus, the kernel of $\mathcal{G}$, denoted by $K(\mathcal{G})$, is the set of all the landmark areas having a non-zero intersection with $\mathcal{G}$. The disks in $K(\mathcal{G})$ are called the *kernel disks*. The other landmark disks are called the *non-kernel disks*.

The *directional preimage* of $\mathcal{G}$, for any given commanded direction of motion d, is the region $P(\mathcal{G}, d)$ defined as the largest subset of the workspace such that, if the robot executes the I-command $(d, K(\mathcal{G}))$ from any position in $P(\mathcal{G}, d)$, then it is guaranteed to reach $K(\mathcal{G})$ and thus to stop in $K(\mathcal{G})$. From the entry point in the kernel, the robot can attain $\mathcal{G}$ by executing a P-command.

Fig. 2 shows an example of a directional preimage: the black disks are obstacles; all other disks are kernel landmark disks.

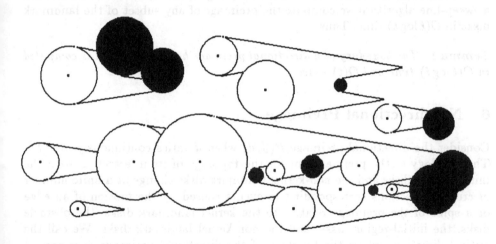

Fig. 2. A directional preimage

$P(\mathcal{G}, d)$ consists of one or several connected subsets bounded by circular segments (*arcs*) and straight segments (*edges*). Each arc is a subset of the boundary of a kernel disk or an obstacle. Let the *left ray* (resp. *right ray*) of a kernel disk L be the half-line tangent to L erected from the tangency point in the direction pointed by $\pi + d + \theta$ (resp. $\pi + d - \theta$), with L on its right-hand side (resp. left-hand side). The *right ray* (resp. *left ray*) of an obstacle disk is defined in the same way, but with orientation $\pi + d + \theta$ (resp. $\pi + d - \theta$), with the obstacle on its right-hand side (resp. left-hand side). Each edge of $P(\mathcal{G}, d)$ is contained in the right or left ray of some kernel disk or obstacle disk. One extremity of the

---

[1] Unlike in Section 2, the kernel is not a subset of the goal region.

edge is the tangency point of the ray; the other extremity is the first intersection point of the ray with another kernel disk, an obstacle disk, or another erected ray. If two edges share the same endpoint, they form a *spike*.

As there are ℓ disks in total, a kernel consists of $O(\ell)$ disks; the number of obstacle disks is also $O(\ell)$. Notice that the arcs on the boundary of the preimage must lie either on the boundary of the union of the kernel disks, or on the boundary of the union of the obstacle disks. We know that the size of the boundary of a collection of disks is linear in the number of disks. Also, each kernel disk and each obstacle disk contributes at most two rays to the preimage, therefore the number of rays is linear. Each ray is interrupted when it hits another ray, a kernel disk, or an obstacle disk, so it cannot contribute more than one additional arc to the boundary of the kernel or the obstacle disks. Therefore, the number of circular arcs on the boundary of a preimage is linear. The number of straight edges on the boundary of a preimage is also linear, because each edge corresponds to one specific ray.

Using a divide and conquer algorithm we precompute all maximal connected sets of landmark or obstacle disks in time $O(\ell \log^2 \ell)$ and space $O(\ell)$. Then, with a sweep-line algorithm we compute the preimage of any subset of the landmark disks in $O(\ell \log \ell)$ time. Thus:

Lemma 1. *The boundary of a directional preimage has size $O(\ell)$. It is computed in $O(\ell \log \ell)$ time and $O(\ell)$ space.*

6 Nondirectional Preimage

Consider the directional preimage $P(\mathcal{G}, d)$ when d varies continuously over S^1. The topology of the preimage and/or the topology of its intersection with the initial-region disks and the non-kernel landmark disks change at a finite number of critical directions corresponding to events caused by the motion of an edge or a spike of the preimage relative to the kernel landmark disks, the obstacle disks, the initial-region disks, and the non-kernel landmark disks. We call the critical directions where the topology of the directional preimage does change the *D-critical* directions and those where the topology of the intersection of the directional preimage with the initial-region disks (resp. the non-kernel landmark disks) changes, other than the D-critical directions, the *I-critical* (resp. *L-critical*) directions. The following description assumes that we move counterclockwisely along S^1.

D-Critical Directions The description of a directional preimage changes when and only when one of the following events occur. Some events are due to kernel disks (we call them K-... events); the others are due to obstacle disks (O-... events).

The K-events are the following (see Fig. 3, where white disks are kernel disks): - A *K-Left-Touch* event occurs when a left edge reaches a kernel disk by becoming tangent to it.

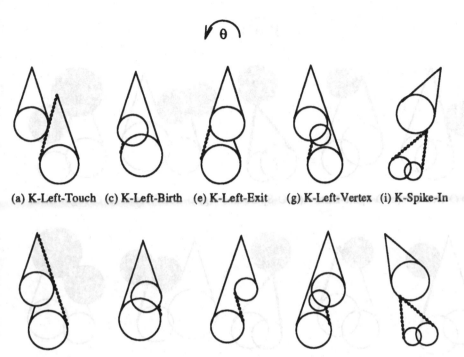

(a) K-Left-Touch   (c) K-Left-Birth   (e) K-Left-Exit   (g) K-Left-Vertex   (i) K-Spike-In

(b) K-Right-Touch   (d) K-Right-Death   (f) K-Right-Exit   (h) K-Right-Vertex   (j) K-Spike-Out

Fig. 3. K-events responsible for D-critical directions

- A *K-Right-Touch* event occurs when a right edge reaches a kernel disk by becoming tangent to it.
- A *K-Left-Birth* event occurs when a new left edge emerges at the intersection of two kernel disks.
- A *K-Right-Death* event occurs when a right edge disappears at the intersection of two kernel disks.
- A *K-Left-Exit* event occurs when a left edge leaves the kernel disk containing its endpoint by becoming tangent to it.
- A *K-Right-Exit* event occurs when a right edge leaves the kernel disk containing its endpoint by becoming tangent to it.
- A *K-Left-Vertex* event occurs when the endpoint of a left edge crosses the intersection between two kernel disks.
- A *K-Right-Vertex* event occurs when a right edge crosses the intersection between two kernel disks.
- A *K-Spike-In* event occurs when a spike reaches a kernel disk.
- A *K-Spike-Out* event occurs when a spike leaves a kernel disk.

The O-events are the following (see Fig. 4, where black disks are obstacle disks):
- An *O-Reach* event occurs when a left edge reaches an obstacle disk by becoming tangent to it.

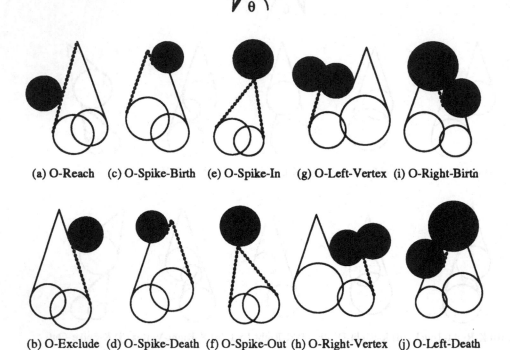

(a) O-Reach (c) O-Spike-Birth (e) O-Spike-In (g) O-Left-Vertex (i) O-Right-Birth

(b) O-Exclude (d) O-Spike-Death (f) O-Spike-Out (h) O-Right-Vertex (j) O-Left-Death

Fig. 4. O-events responsible for D-critical directions

- An *O-Exclude* event occurs when a right edge leaves an obstacle disk by becoming tangent to it.
- An *O-Spike-Birth* event occurs when a spike emerges as a left edge terminating on an obstacle disk reaches the point where a right edge arises from this disk.
- An *O-Spike-Death* event occurs when a spike vanishes as its left edge, pushed by its right edge, shortens to zero length against an obstacle disk.
- An *O-Spike-In* event occurs when a spike enters an obstacle disk.
- An *O-Spike-Out* event occurs when a spike exits an obstacle disk.
- An *O-Left-Vertex* event occurs when the endpoint of a left edge reaches the intersection of two obstacle disks.
- An *O-Right-Vertex* event occurs when the endpoint of a right edge reaches the intersection of two obstacle disks.
- An *O-Right-Birth* event occurs when a right edge emerges at the intersection of two obstacle disks.
- An *O-Left-Death* event occurs when a left edge disappears at the intersection of two obstacle disks.

I-Critical Directions The description of the intersection of $P(\mathcal{G}, d)$ with initial-region disks may change at some D-critical values of d. It also changes at I-critical

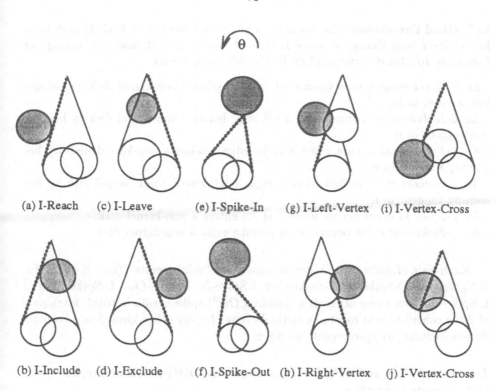

(a) I-Reach   (c) I-Leave   (e) I-Spike-In   (g) I-Left-Vertex   (i) I-Vertex-Cross

(b) I-Include   (d) I-Exclude   (f) I-Spike-Out   (h) I-Right-Vertex   (j) I-Vertex-Cross

Fig. 5. Events responsible for I-critical directions

directions corresponding to the following events (see Fig. 5, where the grey disk is an initial-region disk):
- An *I-Reach* event occurs when a left edge reaches an initial-region disk by becoming tangent to it.
- An *I-Include* event occurs when a left edge leaves an initial-region disk by becoming tangent to it.
- An *I-Leave* event occurs when a right edge reaches an initial-region disk by becoming tangent to it.
- An *I-Exclude* event occurs when a right edge leaves an initial-region disk by becoming tangent to it.
- An *I-Spike-In* event occurs when a spike enters an initial-region disk.
- An *I-Spike-Out* event occurs when a spike exits an initial-region disk.
- An *I-Left-Vertex* event occurs when a left edge crosses the intersection of the kernel disk containing its endpoint with an initial-region disk.
- An *I-Right-Vertex* event occurs when a right edge crosses the intersection of the kernel disk containing its endpoint with an initial-region disk.
- An *I-Vertex-Cross* event occurs when the origin of an edge crosses the intersection of its kernel disk with an initial-region disk.

L-Critical Directions The description of the intersection of $P(\mathcal{G}, d)$ with non-kernel disks may change at some D-critical values of d. It may also change at L-critical directions corresponding to the following events:

- An *L-Reach* event occurs when a left edge reaches a non-kernel disk by becoming tangent to it.
- An *L-Include* event occurs when a left edge leaves a non-kernel disk by becoming tangent to it.
- An *L-Leave* event occurs when a right edge reaches a non-kernel disk by becoming tangent to it.
- An *L-Exclude* event occurs when a right edge leaves a non-kernel disk by becoming tangent to it.
- An *L-Spike-In* event occurs when a spike enters a non-kernel disk.
- An *L-Spike-Out* event occurs when a spike exits a non-kernel disk.

Each pair of disks contributes at most two "spike events" (i.e., K-Spike-In, K-Spike-Out, O-Spike-In, O-Spike-Out, I-Spike-In, I-Spike-Out, L-Spike-In, and L-Spike-Out) for every other disk, yielding $O(\ell^3)$ spike events in total. Each pair of disks contributes at most one critical event of every other kind. Since I-events do not include any spike event, we have:

Lemma 2. *There are $O(\ell^3)$ D-critical directions, $O(\ell^2)$ I-critical directions, and $O(\ell^3)$ L-critical directions.*

Computing the critical directions caused by a spike event requires intersecting a four-degree curve (the curve traced by the corresponding spike) with circles. The analytic solution to this problem is given in [11].

Let us assume that the various disks are in general position, i.e., no two events occur simultaneously.

Let $(d_{c_1}, \ldots, d_{c_p})$ be the cyclic list of all critical directions in counterclockwise order and $I_1, \ldots, I_p$ be the regular intervals between them, with $I_i = (d_{c_i}, d_{c_{i+1(\mathrm{mod}\ p)}})$. For any interval I_i, let d_{nc_i} be any direction in I_i. In order to characterize all the directional preimages of $\mathcal{G}$ and their intersection with $\mathcal{I}$ and the non-kernel disks, it suffices to compute $P(\mathcal{G}, d)$ for all $d \in \{d_{nc_1}, d_{c_1}, d_{nc_2}, \ldots, d_{c_p}\}$. The set $NP(\mathcal{G})$ of all these directional preimages is called the *nondirectional preimage* of $\mathcal{G}$ [6].

To compute the nondirectional preimage we first construct the directional preimage of $\mathcal{G}$ for an arbitrary direction $d_\alpha \in S^1$. The rest of the computation is an alternation of event scheduling and event processing phases (see [11] for more detail).

Sorting the $O(\ell^3)$ critical events takes $O(\ell^3 \log \ell)$ time. Most of them can be processed in constant time, some need $O(\log \ell)$ time, and yet others, called *catastrophic events*, require the recomputation of the directional preimage. The catastrophic events are the K-Left-Touch and K-Right-Exit events, which are only $O(\ell^2)$ in number. Therefore, we have:

Lemma 3. *The computation of the nondirectional preimage takes $O(\ell^3 \log \ell)$ time and $O(\ell^3)$ space.*

7   Planning Method

If the initial region $\mathcal{I}$ is not contained in the goal region $\mathcal{G}_0$, the planner first computes the kernel $K(\mathcal{G}_0)$. If $K(\mathcal{G}_0) = \emptyset$, the planner returns failure. Otherwise, it associates a P-command to reach a goal point with every landmark disk in this kernel. If $\mathcal{I} \subset K(\mathcal{G}_0)$ the planner returns success.

Assume that $\mathcal{I} \not\subset K(\mathcal{G}_0)$. The planner then computes the nondirectional preimage $NP(\mathcal{G}_0)$. If $NP(\mathcal{G}_0)$ contains a directional preimage $P(\mathcal{G}_0, d)$ that includes $\mathcal{I}$, then the planner attaches the I-command $(d, K(\mathcal{G}_0))$ to $\mathcal{I}$ and returns success. Otherwise, for every landmark area $LA \not\subset K(\mathcal{G}_0)$ that has a non-zero intersection with a directional preimage $P(\mathcal{G}_0, d)$ in $NP(\mathcal{G}_0)$, an "exit point" is arbitrarily selected in $LA \cap P(\mathcal{G}_0, d)$ and the I-command $(d, K(\mathcal{G}_0))$ is attached to this point. (If the same area LA intersects several directional preimages, only one intersection is used to produce the I-command.)

The union of the directional preimages in $NP(\mathcal{G}_0)$ is now recursively considered as a goal $\mathcal{G}_1$. Its kernel $K(\mathcal{G}_1)$ is constructed. By construction, $K(\mathcal{G}_1) \supseteq K(\mathcal{G}_0)$. If $K(\mathcal{G}_1) = K(\mathcal{G}_0)$, the planner terminates with failure since it cannot compute a larger nondirectional preimage than $NP(\mathcal{G}_0)$. Otherwise, every landmark area in $K(\mathcal{G}_1) \backslash K(\mathcal{G}_0)$ contains one disk L with an exit point and an I-command attached to it. With every other disk in the landmark area, the planner associates a P-command to reach the exit point in L. If $\mathcal{I} \subset K(\mathcal{G}_1)$ the planner returns success, else it computes the nondirectional preimage of $\mathcal{G}_1$, and so on.

During this backchaining process, the set of landmark areas in the kernels of the successive goals increases monotonically. At every iteration, either there is a new landmark area in the kernel, and the planner proceeds further, or there is no new area, and the planner terminates with failure. The planner terminates with success whenever it has constructed a kernel $K(\mathcal{G}_n)$ containing $\mathcal{I}$ or a nondirectional preimage $NP(\mathcal{G}_n)$ that includes a directional preimage containing $\mathcal{I}$. Let s be the number of landmark areas. The number of iterations is bounded by s. Thus, $n \leq s$. The total time complexity of the planner is $O(s\ell^3 \log \ell)$.

Using our assumptions about the sensing capabilities of the robot and the definition of the preimage, we can prove that our planner does not miss any existing guaranteed plan. Additionally, since we backchain nondirectional preimages, the number of iterations is the lowest possible. Using amortizing techniques described in [11], we can slightly improve the computational complexity of the planner.

Theorem 4. *The planning algorithm is complete and generates optimal guaranteed plans.[2] It takes $O((s + \log \ell)\ell^3)$ time and $O(\ell^3)$ space.*

---

[2] *Here a plan is said to be* optimal *if the maximal number of executed motion commands is minimal over all possible guaranteed plans.*

The computed plan is represented as an unordered set of motion commands distributed over the initial-region disks and landmark disks (see Fig. 1). This kind of representation is reminiscent of the concept of a "reaction plan" introduced in classical AI planning research [17]. If the input problem admits no guaranteed plans, the planner nevertheless constructs such a plan, but with no motion command attached to the initial-region disks. Then the robot can try to enter one of the landmark areas where a command is available, by performing an initial random motion with reflection on the obstacles' boundary. If an unexpected event occurs at execution time, the robot may attempt to reconnect to the plan in the same way.

8 Experimental Results

The above planning method is implemented in C on a DECstation 5000, along with a robot simulator. We experimented with it on many examples. The example shown in Fig. 1 is one of them. It required the computation of 12 successive nondirectional preimages and took about 3.5 minutes of computation time. See [11] for more experimental examples.

9 Discussion and Extensions

We now present extensions of the planner which eliminate or alleviate the most restrictive assumptions contained in the problem statement of Section 3. We expect that most of these extensions can be worked out while keeping the planner polynomial. So far we only have implemented the last extension.

- *Uncertainty in landmark areas:* Perhaps the less realistic assumption in the problem statement of Section 3 is that control and sensing are perfect in landmark areas. However, this assumption can be partially eliminated. Indeed, a landmark area is included in a kernel if it has a non-zero intersection with the goal or a previously computed directional preimage. At execution time, if the robot enters this area, it must only be guaranteed that it can move into this non-zero intersection before it executes the next step in the plan. This does not require perfect sensing or control.

- *Landmark and obstacle geometry:* In our current implementation, the landmark and obstacle areas must be unions of circular disks. But the algorithm can be easily adapted to deal with areas described as generalized polygons bounded by straight and circular edges. It can also accept landmark and obstacles areas that touch each other. The degree of any curve traced by a spike remains no greater than 4. If the workspace contains s landmark areas bounded by $O(\ell)$ edges and arcs and the obstacle areas are bounded by $O(\ell)$ edges and arcs, the time complexity of the planner is $O((s + \log \ell)\ell^3)$. Representing landmark and obstacle areas as generalized polygons is a very realistic model for most applications.

- *Compliant motions:* The planning algorithm can be extended to allow compliant motion commands making the robot slide in obstacle boundaries [14].

This extension would enlarge the set of problems admitting guaranteed plans. In general, it would also allow the planner to produce shorter motion plans.

- *Varying directional uncertainty:* Let $P_\theta(\mathcal{G}, d)$ denote the directional preimage of $\mathcal{G}$ for the direction d computed with the directional uncertainty θ. Let us say that a value θ_c of the directional uncertainty is *critical* if there exists $d \in S^1$ such that the intersections of $P_{\theta_c+\epsilon}(\mathcal{G}_i, d)$ and $P_{\theta_c-\epsilon}(\mathcal{G}_i, d)$, for an arbitrarily small ϵ, with the initial-region and the non-kernel disks are topologically different. The equations defining θ_c can be established by tracking the variations of the nondirectional preimage contour when θ varies. Using this idea, we have recently implemented a polynomial variant of the planner presented in this paper. This variant generates plans for robots that can control directional uncertainty in a given interval, with greater costs assigned to motion commands with smaller uncertainty [12].

10 Conclusion

We described a complete polynomial planning algorithm for mobile-robot navigation in the presence of uncertainty. The algorithm solves a class of problems where landmarks are scattered across the workspace. With the possible exception of [9], previous algorithms to plan motion strategies under uncertainty were either exponential in the size of the input problem, or incomplete, or both. Our work shows that it is possible to identify a restricted, but still realistic, subclass of planning problems that can be solved in polynomial time. This subclass is obtained through assumptions whose satisfaction may require prior engineering of the robot and/or its workspace. In our case, this implies the creation of adequate landmarks, by taking advantage of the natural features of the workspace, or introducing artificial beacons, or using specific sensors. This work gives a formal justification to robot/workspace engineering: make planning problems tractable. Engineering the robot and the workspace has its own cost and we would like to minimize it. Our future research will address this issue.

Acknowledgments: This research has been partially funded by DARPA contract DAAA21-89-C0002 and ONR contract N00014-92-J-1809.

References

1. Briggs, A.J., "An efficient Algorithm for One-Step Planar Compliant Motion Planning with Uncertainty," *Proc. of the 5th Annual ACM Symp. on Computational Geometry*, Saarbruchen, Germany, 1989, pp. 187-196.
2. Buckley, S.J., Planning and teaching Compliant Motion Strategies, Ph.D. Dissertation, Department of Electrical Engineering and Computer Science, MIT, Cambridge, MA (1986).
3. Canny, J.F. and Reif, J., "New Lower Bound Techniques for Robot Motion Planning Problems," *27th IEEE Symp. on Foundations of Computer Science*, Los Angeles, CA, 1987, pp. 49-60.

83

4. Canny, J.F., "On Computability of Fine Motion Plans," *Proc. of the IEEE Int. Conf. on Robotics and Automation*, Scottsdale, AZ, 1989, pp. 177-182.
5. Christiansen, A., Mason, M. and Mitchell, T.M., Learning Reliable Manipulation Strategies without Initial Physical Models, *IEEE Int. Conf. on Robotics and Automation*, Cincinnati, OH (1990).
6. Donald, B.R., "The Complexity of Planar Compliant Motion Planning Under Uncertainty," *Algorithmica*, 5, 1990, pp. 353-382.
7. Donald, B.R. and Jennings, J., "Sensor Interpretation and Task-Directed Planning Using Perceptual Equivalence Classes," *Proc. of the IEEE Int. Conf. on Robotics and Automation*, Sacramento, CA, 1991, pp. 190-197.
8. Erdmann, M., *On Motion Planning with Uncertainty*, Tech. Rep. 810, AI Lab., MIT, Cambridge, MA, 1984.
9. Friedman, J., *Computational Aspects of Compliant Motion Planning*, Ph.D. Dissertation, Report No. STAN-CS-91-1368, Dept. of Computer Science, Stanford University, Stanford, CA, 1991.
10. Latombe, J.C., Lazanas, A., and Shekhar, S., "Robot Motion Planning with Uncertainty in Control and Sensing," *Artificial Intelligence J.*, 52(1), 1991, pp. 1-47.
11. Lazanas, A., and Latombe, J.C., *Landmark-Based Robot Navigation*, Tech. Rep. STAN-CS-92-1428, Dept. of Computer Science, Stanford, CA, 1992.
12. Lazanas, A., and Latombe, J.C., *Motion Planning with Controllable Directional Uncertainty*, Tech. Rep., Dept. of Computer Science, Stanford, CA, 1992.
13. Levitt, T.S., Lawton, D.T., Chelberg, D.M. and Nelson, P.C., "Qualitative Navigation," *Image Understanding Workshop*, Los Angeles, CA, 1987, pp. 447-465.
14. Lozano-Pérez, T., Mason, M.T. and Taylor, R.H., "Automatic Synthesis of Fine-Motion Strategies for Robots," *Int. J. of Robotics Research*, 3(1), 1984, pp. 3-24.
15. Mahadevan, S. and Connell, J., Automatic Programming of Behavior-based Robots using Reinforcement Learning, IBM T.J. Watson Res. Rept. (1990).
16. Natarajan, B.K., The Complexity of Fine Motion Planning, *Int. J. of Robotics Research*, 7(2):36-42, 1988.
17. Schoppers, M.J., *Representation and Automatic Synthesis of Reaction Plans*, Ph.D. Dissertation, Dept. of Computer Science, University of Illinois at Urbana-Champaign, Urbana, IL, 1989.

Using Genetic Algorithms for Robot Motion Planning

Juan Manuel Ahuactzin(i)*, El-Ghazali Talbi(ii) **,
Pierre Bessière(ii), Emmanuel Mazer(i) ***

Institut National Polytechnique de Grenoble
46, Avenue Félix Viallet, 38031 Grenoble cedex - France
Tel: (33) 76.57.48.13 Fax: (33) 76.57.46.02

Abstract. We present an ongoing research work on robot motion planning using genetic algorithms. Our goal is to use this technique to build fast motion planners for robot with six or more degree of freedom. After a short review of the existing methods, we will introduce the genetic algorithms by showing how they can be used to solve the invers kinematic problem. In the second part of the paper, we show that the path planning problem can be expressed as an optimization problem and thus solved with a genetic algorithm. We illustrate the approach by building a path planner for a planar arm with two degree of freedom, then we demonstrate the validity of the method by planning paths for an holonomic mobile robot. Finally we describe an implementation of the selected genetic algorithm on a massively parallel machine and show that fast planning response is made possible by using this approach. Keywords: robot motion planning, genetic algorithms, parallel algorithms.

1 Introduction

Today most of the robot motion planners are used offline: the planner is invoked with a model of the environment, it produces a path which is passed to the robot controller which, in turn, execute it. In general, the time necessary to achieve this loop is not short enough to allow the robot to move in a dynamical environment. Our goal is to try to reduce this time to be able to move a six degree of freedom arm among moving obstacles. In order to achieve this goal we have chosen genetic algorithms for the following reasons:

---

* (i) Laboratoire d'Informatique Fondamentale et d'Intelligence Artificielle Grenoble, France.
** (ii) Laboratoire de Genie Informatique, IMAG, Grenoble, France.
*** This work has been made possible by: Le Centre National de la Recherche Scientifique, Consejo Nacional de Ciencia y Tecnologia (Mexico) and ESPRIT "Supernode 2" project.

- They are well adapted to search for solutions in high dimensionality search space.
- They are very tolerant to the form of the function to optimize, for instance these functions do not need to be neither differentiable or continuous.
- They can easily be implemented on a massively parallel machine and they achieve super-linear speed up with the number of processors [8].

In addition we have tried to take advantage of tools already existing in classical algorithms used in path planning, such as range computation, slice representation of the configuration space and on demand computation of the configuration space.

2   Previous work

Designing a new path planner is a classical exercise in robotic research and it remains a very active field in robotics. A review of the existing approaches can be found in Latombe's book [1]. Recently a real time path planner system has been demonstrated by [2]. The method works for a three degree of freedom robot assuming the three dimensional configuration space has been precomputed. The system described in [3] is probably one of the fastest system if one consider the time necessary to compute the configuration space and the time necessary to search through it. It uses an efficient way of computing the configuration space and it is implemented on a Connection Machine.

3   Genetic algorithms

Genetic algorithms are stochastic search techniques, introduced by Holland [4] twenty years ago, they are inspired by adaptation in evolving natural systems. Development of massively parallel architectures made them very popular in the very last years. To introduce the genetic algorithms and their application in path planning, we will first solve a simple robotic problem: the invers kinematic problem.

3.1   Solving the invers kinematic problem

Lets consider an arm with two degree of freedom an lets denote $\theta = (\theta_1, \theta_2)$ a particular configuration of the arm. One version of the invers kinematic problem can be stated as a minimization problem:

$$\min_{\theta} f(\theta) \ with \ f(\theta) = \begin{cases} \|d(\theta) - X\| & \text{if } \theta \notin \text{C-obstacles} \\ +\infty & \text{otherwise} \end{cases}$$

Where $d(\theta)$ denotes the direct kinematic function and X the desired cartesian location for the extremity of the arm.

In general, it is very hard to get an analytical expression of f given a set of obstacles, but computing its value for a given θ can be done easily.

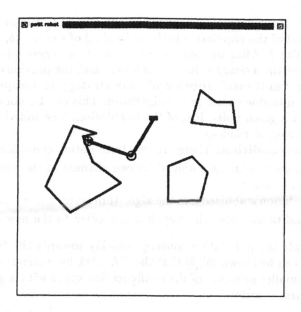

Fig. 1. A planar arm with two degrees of freedom.

In the rest of this section we show how to use GA to solve the invers kinematic problem with obstacles.

1. **Coding the problem.** The search space (here $[0, 2\pi] \times [0, 2\pi]$) is discretized: In our case a configuration is represented by two integers (i_1, i_2), where i_1 and i_2 belong to $0, \ldots, 255$. We denote as b the string of bits resulting from the concatenation of the binary representation of i_1 and i_2, for example if $i_1 = 1$ and $i_2 = 2$: $b = 00000010000010$. b codes a point of the discretized search space. In the GA terminology b is referred as a "individual".

2. **Generating an initial population.** A random set of n "individuals" $B = \{b_i\}_{i=1,n}$ is generated. For example figure 1 represents our small robot in a clustered environment and figure 2 the associated configuration space. The small • indicates a particular $b_j \in B$, the ∘ sign indicates the desired locations to place the extremity of the arm in the desired cartesian position.

3. **Operating a selection.** The function f is applied to each member of the population and the "individuals" are ranked accordingly to their associated value of f. After this step, the closer a configuration brings the extremity of the arm to the desired goal without collision, the better its rank will be.

4. **Creating couples and combining individuals.** A set of n couples $\{(b_i, b_j)\}$, $b_i, b_j \in B$ is generated. For each couple, b_i and b_j are obtained by randomly picking an element of B with a probability proportional to their rank.
   For each couple the "cross-over" operation is used to produce two new individuals nb_1 and nb_2. This operation consists of generating at random a place to "cut" the binary strings which codes the two configurations defining the couple, to "glue" the left part of b_i with the right part of b_j to obtain nb_1

and to "glue" the right part of b_i with the left part of b_j to obtain nb_2. The new individual of the population is the individual of the set $\{b_i, b_j, nb_i, nb_j\}$ which minimize f. After this step a new population is generated. It can be shown [5] that the average value of f is lower than the previous population.

5. **Generating "mutants" (optional)** One strategy to escape from local minima is to introduce noise in the algorithm. This can be done by flipping at a given bit a given individual of the population. The individual and the bit bit are chosen at random.

6. **Termination conditions** There are two termination conditions:

   (a) The absolute minima is obtained for one element of the population and a solution is found.

   (b) The population stabilized and the algorithm is stuck in a local minimum.

   If neither conditions is true the step # 3 is applied to the new population.

In our example the population converge quickly towards the two solutions (see figure 3). It can be shown [5],[6] that the GA work by interratively selecting an increasingly smaller subspace of the configuration space with a good average value for the fitness function.

4 A simple path planner

In this section we describe a simple path planner for a planar arm with two degrees of freedom. By restricting ourself to two dimensions we can graphically represent the configuration space and give the reader a better feeling of the method. However the proposed method does not make any hypothesis about the number of degree of freedom and can be used without modification for arms with having a larger number of degree of freedom.

To plan a path with a GA we use a quite unexpected search space. Instead of searching for path in the configuration space, we consider a discretized subset of all the possible paths (with or without collision) starting from the initial configuration. For example, in the case of our planar robot we consider the following subset of possible motions: "move from θ_1^0, move from θ_2^1, move from θ_1^2, move from θ_2^3, ... move from θ_1^i, move from θ_2^{i+1}, ..., θ_1^{2k}, move from θ_2^{2k+1}" where each θ_i^j is discretized on q bits.

We call such a motion a Manhattan motion of length k and we denote the set of all these paths as $M(k, q)$. Figure 4 shows such a motion in the configuration space.

They are four main reasons to consider Manhattan motions:

1. They define a naturally redundant search space which is well suited for stochastic methods.

2. They are easy to test: When a Manhattan motion is executed only one link move at a time, and thus the test can be done by successively computing the legal ranges of motion for a single degree of freedom [7] . Note that this last property remains true for three dimensional objects and for robot with many degree of freedom.

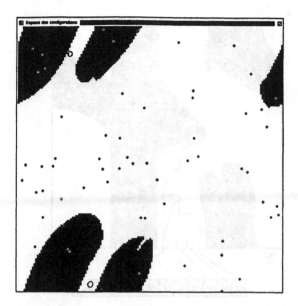

Fig. 2. Initial distribution of the population in the configuration space .

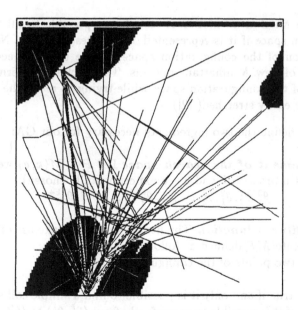

Fig. 3. The population converges twoards the two solutions of the invers kinematic problem.

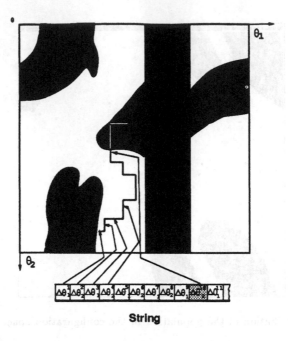

String

Fig. 4. A Manhattan motion in the configuration space.

3. Each time a Manhattan motion is tested it is possible to fill a part of the configuration space if it is represented with slices(see [7]). Note that this representation of the configuration space can be use to speed up further evaluations of new Manhattan motions. This technique permits the lazy evaluation of the configuration space while trying to reach the goal.
4. They can be easily stretched([2])

We can now define the two ingredients necessary for our GA:

Coding the element of the search space Let $P \in M(k, q)$ we can code P with $k \times q$ bits (in practice k and q are experimentally chosen):
$$\theta_1^0, \theta_2^1, \theta_1^2, \theta_2^3, \ldots, \theta_1^i, \theta_2^{i+1}, \ldots, \theta_1^{2k}, \theta_2^{2k+1}$$

Defining the fitness function The fitness function make use of a special accessibility predicate MP defined as follow.

Let I and G two points of the configuration space:

$MP(I, G)$ is true if and only it is possible to find a simple collision free path from I to G (ie: if it is possible to move freely from (θ_1^I, θ_2^I) to (θ_1^G, θ_2^I) and from (θ_1^G, θ_2^I) to (θ_1^G, θ_2^G).

90

Note that the definition of this predicate can easily be evaluated and generalized. Now, we can define how to compute the fitness function:

Let G be the goal and S_i be the extremity of the ith segment of the Manhattan path P:

```
begin
for i from 0 to 2k
    if MP(S_i, G) return 0
    else if ¬ MP(S_i, S_{i+1}) return ||S_i - G||
        endif
    endif
endfor
return ||S_{2k+1} - G||
end
```

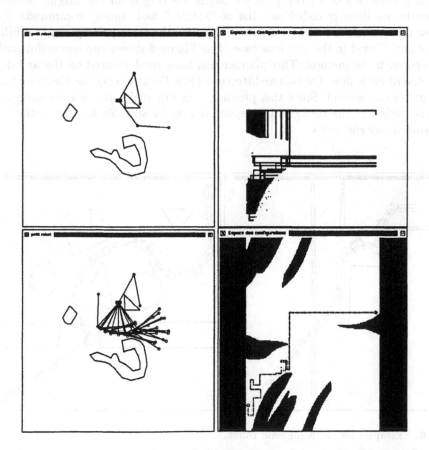

Fig. 5. A motion planning problem and the solution in the operational and configuration spaces.

This function simply scan the intermediary points of the candidate path to find if a "direct" move towards the goal, it stops when an obstacle is found and return the distance from the extremity of the last free segment to the goal.

Figure 5 shows the initial and the final position of the robot in the operational space. At the botton of the figure we show a path found with this method in the two spaces. The right top of the figure is the portion of the configuration space which has been evaluated during the planning process, in this particular case only 1/10 of the total configuration space has been evaluated.

5   Planning a path for an holonomic mobile robot

In this section we describe the application of our algorithm to the problem of planning a collision free path for an holonomic mobile robot. The search space is still a subspace of all the possible paths starting from the origin, however the paths are directly coded as a list of "rotate " and "move" commands. The fitness function is also slightly different but uses the same principle of "visibility to the goal " used in the previous case. The Figure 6 shows two successful paths planned with the method. This planner has been implemented on the architecture described bellow. On this architecture (128 Transputers) the planning time was under one second. Since this planner does not make use of a precomputed representation of the configuration space it may be suitable to plan paths in a dynamical environments.

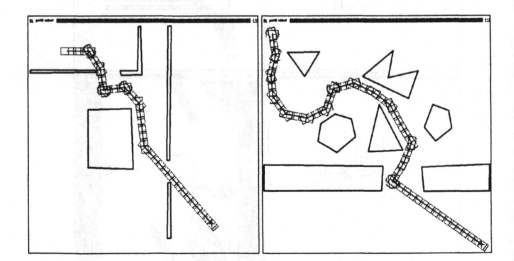

Fig. 6. Examples for the holonomic robot.

6 Parallel Genetic Algorithm

Besides the "intrinsic parallelism" of genetic Algorithms [4][5], we designed and implemented a parallel genetic algorithm which shows a remarkable super-linear speed-up in the number of processors [8]. The principle of this algorithm is to distribute the individuals of the population on a set of processors, limiting the reproduction (step 3 and 5 of the standard algorithm) to the nearby individuals. The implementation has been done on a SuperNode [8]c.

The SuperNode is a loosely coupled and highly parallel machine based on transputers. One of its most important characteristics is its ability to dynamically reconfigure the network topology by using a programmable VLSI switch device. This architecture offers a range of 16 to 1024 processors, delivering from 24 to 1500 Mflops performance. The adopted configuration of the machine is a torus. Given the four links of the transputer, each individual has four neighbors (see figure 7). The distribution of the population on the set of processors has been done on a one individual by processor basis. Consequently, at each generation on individual is susceptible reproduce only with its four direct neighbors. The parallel genetic algorithm is as follow:

- **Step 1** Distribute and generate the initial population on the set of processors (one individual by processor).
- **Step 2** Evaluate the initial population **IN PARALLEL**.
- **Step 3** Until the stop conditions are reached do **IN PARALLEL**:
  - Exchange with the four neighbors the individuals (chain of bits) .
  - Do the four possible individual combinations (Cross over), between the local individual and each of the four just acquired neighboring individuals.
  - Do (optionally) the mutation.
  - Select the best of the generated offspring and replace the current local individual.

Note that, unlike for the sequential Genetic Algorithm, we are not able to get the best solution in the network. The communication involved in determining such a solution would be considerable. we only pick up the solution routing through a "spy process" placed at the root processor (see figure 7).

This parallel genetic algorithm and its associated implementation have several important qualities:

- The needed configuration of the network of processors (a torus) is very simple and easy to obtain.
- Communication between processors are only local (limited to the four neighbors) and thus supposes only a very simple message routing capability.
- Due to the strictly local form of the message passing strategy, the amount of communications grows linearly with the number of processors. This proves that huge configuration counting hundreds of processors will still permit a valid implementation of the proposed algorithm.

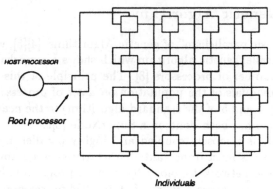

Fig. 7. A torus of 16 processors

7 Future work

We plan to integrate this planner in a larger robotic experimentation tested. Our goal is to use real CAD models to evaluate the fitness function. The CAD models of the arms and of the obstacles will be obtained from the ACT [9] simulation package. The range computation used to evaluate the fitness function will use the same algorithm found in the global path planner of ACT. Our goal is to be able to execute the computed trajectory via KALI [10] and to use the second arm as a dynamical obstacle.

References

1. Jean-Claude Latombe: *Robot Motion Planning*, Ed. Kluwer Academic Publisher, 1991.
2. Sean Quinlan, Oussama Khatib: *Towards Real-Time Execution of Motion Tasks*, Second International Symposium on Experimental Robotics, Toulouse, June 1991.
3. Tomás Lozano-Pérez, Patrick A. O'Donnell: *Parallel Robot Motion Planning*, IEEE Int. Conf. on Robotics and Automation, Sacramento, April 1991.
4. J.H. Holland: *Adaptation in Natural and Artificial Systems*, Ann Arbor: University of Michigan Pres, 1975.
5. David E. Goldberg: *Genetic algorithms in search, optimization and machine learning*, The University of Alabama, Addison-Wesley publishing company, inc. 1989.
6. Davis E. Goldberg: *Genetic algorithms and machine learning*, University of Alabama, Tuscaloosa, University of Michigan, Ann Arbor, in Machine Learning, Kluwer Academic Publishers, 1988.
7. T. Lozano-Pérez: *A simple motion-planning algorithm for general robot manipulators.*, IEEE Int. Jour. on Robotics and Automation, RA-3, June 1987.
8. E.G. Talbi, P. Bessiere: *A parallel Genetic Algorithm for the graph partioning problem*, ACM International Conference on Supercomputing, Cologne, June 1991.
9. Emmanuel Mazer and al. : *ACT a robot programming environment*, IEEE Int. Conf. on Robotics and Automation, Sacramento, April 1991.
10. V. Hayward, L. Daneshmend, S. Hayati. *An overview of KALI: A system to program and control cooperative manipulators*, Fourth international conference on advanced robotics, 1988.

Fast Mobile Robots in Unstructured Environments

René Zapata, P. Lépinay

Laboratoire d'Informatique, de Robotique et de Microélectronique de
Montpellier
UM C9928
Université de Montpellier II
Montpellier, FRANCE

Abstract : This paper addresses the problem of mobile robot reactive behaviors evolving in unstructured and dynamic environments. The emphasis is put on a software structure devoted to obstacle avoidance when geometric models for obstacles do not exis. The first part concerns the model of the robot/environment interaction. This model is based on the definition of the *interaction state* of the robot and leads to avoidance–oriented control laws. The second part describes the implementation of this algorithm on a fast outdoor mobile robot called SNAKE.

1.Introduction

The Motion Planning problem for mobile robots has been thoroughly investigated in the case of structured environments [1,2,3,4,5] .
For a mobile robot, moving amongst unknown obstacles induces the necessity of taking unscheduled and dynamic events into account and trying to react as fast as living beings do.
Therefore, reactive behaviors play a fundamental role when the robot has to evolve in unstructured and dynamic environments.
A few methods have been developed and implemented in order to provide mobile mechanisms with this capability of reaction. As a generic example, we can point out Brooks' work on the subject [6,7]. His approach consists in dividing the robot hardware and software into elementary procedures relating dedicated information to dedicated actions. Each of these modules, called behaviors, allows reactivity in front of exterior stimuli. This general method, based on an ethological view of robotics, has been followed by other researchers [8,9,10].
From another point of view, reactive behaviors can be considered as particular cases of sensor–based control loops. Exteroceptive information, imbedded into the robot control loop, allow direct interactions between the robot and its environment [11]. An interesting method, developed by Holenstein and Bradreddin, is based on the computation of a robot–referred artificial potential field [12]. Local artificial forces, acting on the vehicle, are obtained from ultrasonic information and this, without building any model of the environment.

When fast collision avoidance procedures are unnecessary, we can assert that there are no needs of pattern recognition and object interpretation. The fact that there is a stone or a tree on the robot way is less important than the existence and the global shape of this obstacle. We have developped the concept of *interaction state*, which represents the robot internal representation of its surrounding space without any consideration on the nature of the obstructing objects.

2. A Model of the Robot/Environment Interaction

The robot/environment interaction can be described as a deformable virtual zone (DVZ) surrounding the robot [13]. The deformations of this zone are due to the intrusion of proximity information in the robot space and control the robot reactions. The *robot internal state* is defined to be a couple $(\pi, \varXi)$ where the first component π characterizes the robot dynamics (its translational and rotational velocities) and the second component $\varXi$, called the *interaction component*, characterizes the geometry of this deformable zone. The *internal control* –or reactive behavior– is a relation ϕ, linking these two components, namely $\varXi = \phi(\pi)$.

This means that, the interaction zone, disturbed by obstacle intrusion, can be reformed by acting on the robot controls.

In the following, n will denote the dimension of the robot world (its operational spaxe), $\mathbb{R}$ the real line, $||y||$ the euclidean norm of vector y and $\dfrac{\partial f}{\partial x}$ the jacobian of the vector–valued function f.

2.1.The Interaction Component

The geometric interaction between the moving n–dimensional robot and its moving n–dimensional environment, it is to say the deformable zone surrounding the vehicle, can be viewed as an imbedding of the (n–1)–dimensional sphere S^{n-1} into the cartesian n–dimensional space $\mathbb{R}^n$. Let us recall [14] that an imbedding f of the manifold V into the manifold W is a regular homeomorphism from V into f(V). In other words :

> *# f is one–to–one,*
>
> *# f is regular* $(\forall x \in V$, $rank[\dfrac{\partial f}{\partial x}(x)] = \dim V)$,
>
> *# f and its inverse f^{-1} are continuous.*

The main interest in the use of this formalism lies in the fact that each imbedding of S^{n-1} can continuously be transformed into another imbedding of S^{n-1} . Thus, the deformations of the the interaction zone, due to the intrusion of obstacles in the robot workspace or to the modifications of the robot dynamics π (through the relation $\varXi = \phi(\pi)$) lead to the same mathematical entity (the imbedding).

2.2.Classes and Deformations of Imbeddings

The mathematical operators transforming any imbedding of the sphere into another one, can be grouped into three main families :

The group of diffeomorphisms, denoted Diff(n), which represents the general deformations of S^{n-1},

The affine group denoted A(n), which is the set product of the cartesian space $\mathbb{R}^n$ and the linear transformation group SL(n) (the special linear group of non singular $n \times n$ matrices),

The group of rigid motions, denoted D(n) which is the set product of $\mathbb{R}^n$ and SO(n) (the special orthogonal groups elements of which are the $n \times n$ orthogonal matrices).

These *Deformation Operators* transform an imbedding of S^{n-1} into another one and thus, allow to define classes of imbeddings : two imbeddings $\varXi$ and $\varXi'$ are said to be equivalent with respect to a group G of transformations, and we note : $\varXi \underset{G}{\sim} \varXi'$, if and only if we have :

(1) $\exists \gamma \in G$ *such that* $\varXi' = \gamma \ o \ \varXi$

Actually, the Lie Group structure of the three transformation groups previously defined, makes that the relation $\underset{G}{\sim}$ is an equivalence relation.

If G=Diff(n), 'deformation operator' has to be taken in its usual sense. Each single imbedding is a class in itself for the relation $\underset{G}{\sim}$.

If G=A(n), the deformation operators are homothetic and translational functions of S^{n-1}. There are as many equivalence classes as different shapes of the image $\varXi(S^{n-1})$ (in the usual sense of the term).

The operators of G=D(n) are the usual rigid motions. Two imbeddings $\varXi$ and $\varXi'$ are in different classes if the shape or size of the images $\varXi(S^{n-1})$ and $\varXi'(S^{n-1})$ differ. Figure 1 shows different cases of the 1–sphere deformations.

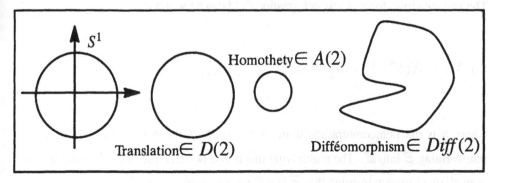

Figure 1 : Deformations of the sphere (circle) S^1

2.3.Deformations and Perception

The relevant information for collision avoidance purposes is certainly the distance polar function relating the closest impact point distance in each direction to this direction of observation.
If we consider that each direction can be observed from the robot point of view, or in other words that the robot is surrounded by a continuous set of virtual proximity sensors, this information can also be represented by an imbedding **B** of the (n–1)–sphere. We will call it the *information boundary manifold*.

What could actually be considered as deformed, is the *range manifold* **R** of all the virtual sensors observing in all the n–dimensional space directions. This continuous set is formed by the ranges of the virtual sensors in all the space directions.

The *intrusion vector field* I is defined on **R** by the vector difference between the information boundary in each direction, and the range of the virtual sensor in the same direction :

(2) $\quad I : r \rightarrow I_r$ with $I_r = \beta_w - r$

where w is the direction of the point r on **R** and β_w the point on the information boundary **B** in this direction.
As we can relate the information intrusion to an imbedding of the sphere, it is theoretically easy to infer an expression of the intrinsic state deformation from Ξ to another interaction component $\tilde{\Xi}$.

This is done by defining a *deformation vector field* Δ from the intrusion vector field. The general close–form of this deformation field can be written :

(3) $\quad \forall \xi \in \Xi(S^{n-1})$, $\Delta_\xi = \alpha(\xi)\, I_{h^{-1}o\Xi^{-1}(\xi)}$

where h is the homeomorphism from S^{n-1} to **R**, Δ the deformation vector field transforming Ξ into $\tilde{\Xi}$. The multiplying function α between the two vector fields, is a normalizing function insuring that $\tilde{\Xi}$ is still an imbedding of S^{n-1}.

Figure 2 summarizes these different concepts for the 2–dimensional world.

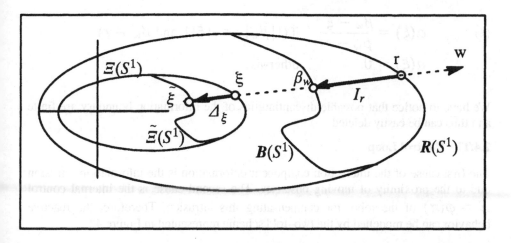

Figure 2 : The Intrusion Vector Field and the induced Deformation Vector Field

The interaction component $\varXi$ can be chosen as a deadband value for the information boundary **B**. In this case, $\varXi$ is deformed if and only if the information boundary manifold **B** is 'smaller' than $\varXi$ (Figure 3). Thus, the normalizing function α works as a trigger.

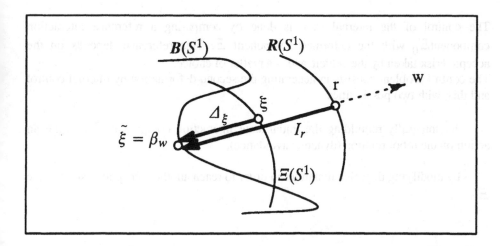

Figure 3 : Deformation by comparing **B** and $\varXi$

From relations (2) and (3) the formal expression of α can be written :

$$(4) \qquad \alpha(\xi) = \frac{\beta_w - \xi}{\beta_w - r} \qquad \text{if } (||\beta_w|| < ||\xi|| \text{ and } \beta_w \neq r)$$

$$\alpha(\xi) = 0 \qquad \text{otherwise}$$

We have to notice that possible discontinuities of the information boundary are finite and thus can be easily deleted.

2.4. The Control Loop

The first cause of the interaction component deformation is the information intrusion due to the proximity of moving obstacles. The second cause is the internal control $\Xi = \phi(\pi)$ of the robot for compensating this intrusion. Therefore, the reactive behavior can be modelled by the two–fold scheme represented in Figure 4 :

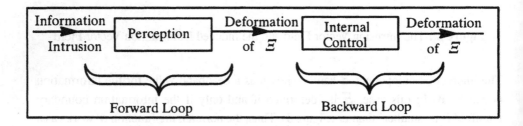

Figure 4 : The control scheme

The control of the internal state is done by comparing a reference interaction component Ξ_0 with the deformed component Ξ. This reference depends on the accepted risk taken by the vehicle and is a matter of choice.

The control problem consists in generating the second deformation by internal control and this, with two possibilities :

– by integrally rebuilding the initial state interaction component Ξ through an action on the robot rotation (dynamic avoidance),

– by modifying the robot dynamics in order to reach another acceptable stable state Ξ'.

3. SNAKE

3.1. Robot description

The reactive behavior structure has been implemented on a fast (5 to 10m/s) outdoor mobile robot. SNAKE (Sensor–based Navigation Autonomous Kinetic Expert) is a

1/4–scale car–like vehicle equipped with an optical velocity sensor, an electronic orientation sensor (magnetic compass) and a 7–sensor ultrasonic system. These regularly spaced sensors cover the robot front world from –45° to 45°. The main motion is provided by an internal combustion engine driven by a DC servo–motor controlling the acceleration cable. Another servo–motor drives the brakes and two other electric motors control the robot direction.

The software has been developed in C language on a Motorola™ 68020/VME architecture. Input/Output boards and timers for controlling sensors and motors complete the structure. SNAKE can work either in an autonomous mode (the robot takes its navigation in charge) or in a remotely operated mode (the human operator remotely guides the vehicle which only has to take care of obstacles).

3.2. Reactive behaviors

Four modules relating sensors to actuators allow an hierarchical description of behaviors (Figure 5). These modules are parallel asynchronous processes.

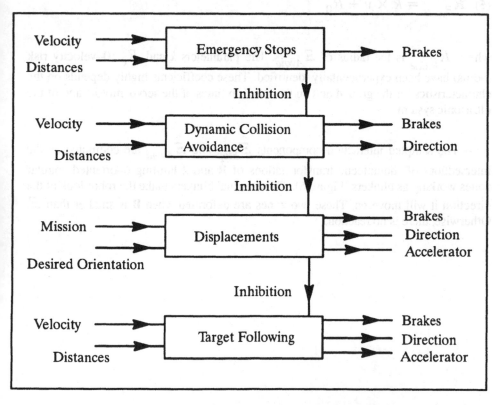

Figure 5 : Control architecture hierarchized by priority

The two low level higt priority modules (Emergency Stops and Dynamic Collision Avoidance) and the high level low priority (Target Following) use the model of the robot internal state (π, Ξ) as follows :

– The robot dynamics π is given by its translational velocity v and its rotational velocity $\dot{\Theta}$. These ones are directly related to the inter–axle distance L and to the turn angle δ of the robot by : $\dot{\Theta} = \dfrac{v}{L} \times \tan(\delta)$.

– The range manifold **R** is an arc of circle and can be deformed along the 7 directions into the information boundary manifold **B**.

– Two circular interaction zones (Ξ_{Stop} for the Emergency Stops procedure and Ξ_{Avoid} for the Dynamic Collision Avoidance procedure) are sampled by using the proximity data provided by the 7 ultrasonic sensors. The sizes of Ξ_{Stop} and Ξ_{Avoid} are proportional to the translational velocity v. For Ξ_{Avoid} we have :

$$(5) \quad R_{\Xi_{Avoid}} = k \times v + R_0$$

where $R_{\Xi_{Avoid}}$ is the radius of Ξ_{Avoid}. The parameters k and R_0 (0–velocity risk radius) have been experimentally identified. These coefficients highly depends on the characteristics of the ground and on the reaction times of the servo–motors and of the ultrasonic system.

– The sampled interaction components $\hat{\Xi}_{Stop}$ and $\hat{\Xi}_{Avoid}$ are computed as the intersections of homothetic transformations of **R** and 2 limiting δ–oriented angular zones working as blinkers (Figure 6). These virtual blinkers make the robot look at the direction it will move on. These two zones are deformed when **B** is smaller than Ξ. Otherwise, there is no reaction.

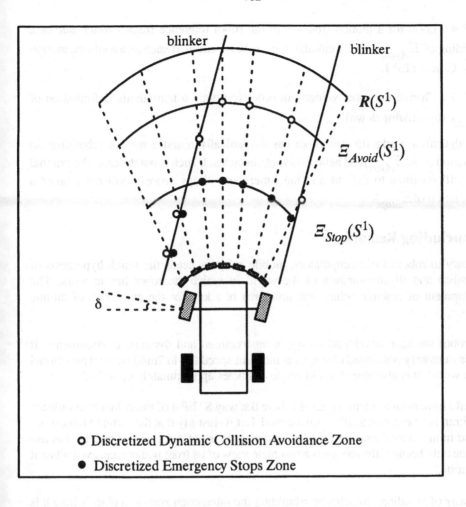

Figure 6 : Discrete Risk Zones

3.3.The Collision Avoidance Algorithm

The internal security zone Ξ_{Stop}(Emergency Stops) is just a triggering space which produces a complete braking of the robot. The use of the external security zone Ξ_{Avoid} is more sophisticated and leads to the following algorithm :

STEP 1 : Check the deformation of **R**. If there is no deformation: Exit.

STEP 2 : Compute Ξ_{Avoid} with respect to the velocity of the robot and its rotational speed.

STEP 3 : Check the deformation of Ξ_{Avoid}. If there is not such a deformation: Exit.

STEP 4 : Look for a motion (rotation in the robot reference frame) which allows a rebuilding of Ξ_{Avoid} while minimizing the robot rotation. If such an avoidance motion exists: Goto STEP 1.

STEP 5 : Change the robot velocity in order to obtain a homothetic deformation of Ξ_{Avoid} (by slowing down).

This algorithm works till the robot has dynamically found a way of rebuilding an interaction zone Ξ_{Avoid} and hence to avoid obstacles. If such is not the case, the internal state will continue to deform and the Emergency Stops process will react (after a deformation of Ξ_{Stop}).

4.Concluding Remarks

Contrary to robotics of manipulators, mobile robotics needs the tough hypothesis of dynamism and ill–structuration of the world in which the robot has to work. The development of reactive behavioral structures is a key for the autonomy of mobile robots.

The robot we have developed moves in unstructured and dynamic environments. It avoids randomly positioned obstacles at medium speed (up to 2m/s) without any model of the world. It is also able to avoid single obstacles approximately up to 7m/s.

It would have been too long to develop here the way STEP 4 of the collision avoidance algorithm has been practically implemented. Let us just say that the virtual blinkers are rotated in the robot frame until the interaction zone is not deformed anymore. This can be done only because the robot has a complete view of its front workspace, even when it is rotating.

This way of avoiding obstacles by rebuilding the interaction zone in a place where it is not deformed anymore could be called 'geometric placement'. Another way could be a 'differential approach' of the problem, by directly feeding back the deformation of the interaction zone (geometry and energy) to the robot control loop, in order to act on the control law.
An interesting approach we are now developping, consists in using Differential Games Theory for the design of action/reaction control laws. In this approach, the robot and the unknown environment are considered as players optimizing the same cost function (minimizing for the first one and maximizing for the other). This cost function is directly related to the deformation of the deformable Virtual Zone (Surface, Energy of Deformation, Time of Deformation,...)

References

1.   L. Bochereau, D. Wolfstein, Y. Kanayama
        "Simulation of model based path planning for a mobile robot",
        9th IASTED International Symposium on Robotics and Auto-
        mation, May 1987, Santa Barbara, CA.

2.   R.A. Brooks
        "Solving the Find–Path Problem by Good Representation of
        Free Space", IEEE Trans. on Sys. Man and Cyb., Vol SMC 13,
        No 3, March April 1983.

3.   J. Canny
        "The Complexity of Robot Motion Planning", PhD. Disserta-
        tion, MIT Press, 1987.

4.   R. Chatila
        "Système de navigation pour un robot mobile autonome :
        modélisation et processus décisionnel", Thèse de docteur ing-
        énieur, Université Paul Sabatier, Toulouse, 1981.

5.   D.E. Koditschek
        "Exact Robot Navigation by Means of Potential Functions :
        Some Topological Considerations", Proc. IEEE Int. Conf. on
        Rob. and Aut., Rayleigh, North Carolina, March 1987, pp. 1–6.

6.   R.A. Brooks
        "Robot Beings", International Workshop on Intelligent Robots
        and Systems '89 (IROS'89), September 1989, Tsukuba, Japon

7.   A. Flynn, R.A. Brooks
        "Building Robots : Expectations and Experiences", Interna-
        tional Workshop on Intelligent Robots and Systems '89
        (IROS'89), September 1989, Tsukuba, Japon .

8.   T.L. Anderson
        "Autonomous Robots and Emergent behaviors : A Set of Prim-
        itives Behaviors for Mobile Robot Control", Proc. IROS'90,
        Tsukuba 1990, Japan

9.   M. Soldo
"Reactive and Preplanned Control in a Mobile Robot", Proc. IEEE Int. Conf. on Rob. and Aut.,Cincinnati, Ohio, USA, May 1990

10.   D. Lawton
"Qualitative Spatial understanding and Reactive Control for Autonomous Robot", Proc. IROS'90, Tsukuba 1990, Japan

11.   B. Espiau, F. Chaumette, P. Rives
"Nouvelle approche de la relation vision–commande en robotique", rapport INRIA RR–1172, 1990

12.   A. Holenstein, E. Badreddin
"Collision Avoidance in a Behavior–based Mobile Robot Design", Proc. IEEE Int. Conf. on Rob. and Aut., Sacramento, California, April 1991.

13.   R. Zapata
"Quelques aspects topologiques de la planification de mouvements et des actions réflexes en robotique mobile", Thèse d'Etat, Université de Montpellier II, Montpellier, juillet 1991.

14.   P. Malliavin
"Géométrie différentielle intrinsèque", Hermann, 1972.

A New Approach to Visual Servoing in Robotics

Bernard Espiau[1], François Chaumette[2], Patrick Rives[3]

[1] ISIA-ENSMP, rue Claude Daunesse, Sophia Antipolis, 06565 Valbonne, France.
[2] IRISA / INRIA Rennes, campus de Beaulieu, 35042 Rennes-cedex, France.
[3] INRIA Sophia Antipolis, 2004 Route des Lucioles, 06565 Valbonne, France.

Abstract. This paper describes a new approach of vision-based control in robotics. The basic idea consists in considering a vision system as a specific sensor dedicated to a task, and included in a control servo-loop. Once the necessary modeling stage is performed, the framework becomes the one of automatic control, and naturally stability and robustness questions arise.

The paper is organized as follows: in the introduction, a short state-of-the-art in the area of visual servoing is reviewed. Then, the basic concepts allowing to model the concerned interactions are given. The *Interaction Screw* is thus defined in a general way, and the application to images follows. Starting from the concept of task function, the general framework of the control is then described, and stability results are recalled. The concept of hybrid task is also presented and then applied to visual sensors.

The paper ends with the presentation of several simulation and experimental results, and some guidelines for future work are drawn in the conclusion.

1 Introduction

How to use vision has always been a major research area in robotics. Early studies in this field, in the 70's, were mainly motivated by problems of pattern recognition. The obtained results together with the huge improvement of the available computing power lead us to consider that the recognition, the localization or the inspection of a motionless part are now in today's state-of-the-art. In parallel, and partly owing to the attention paid to mobile robots, researches have investigated more complex questions: stereo vision, outdoor scene analysis,... In complement, major works were done in the domain of the analysis of image sequences (dynamic vision) [4]. The original motivations came from telecommunication (motion-compensated coding) or military (target tracking, moving objects recognition) applications. Associated algorithms were issued from the signal processing area, and rapidly allowed to consider the possibility of a 'real-time' implementation, i.e. of working at the video rate.

If we now return to the domain of robotics, a recent trend is to use exteroceptive non-contact sensors inside the control servo-loops themselves, and not only as sources of data used in higher decision levels. Such an approach may prove to be very useful if it is necessary, for example, to compensate for small positioning errors, to grasp objects moving on a conveyor belt, to track a seam in arc welding or, more generally, to be adaptable to uncertainties of the environment. Applications of this kind were realized in manipulation robotics as well as in teleoperation or mobile robotics. However, the sensors were mainly optical proximeters [3], [8], [10], [12], [28], or acoustic range finders [2], [16].

In fact, it is also possible to consider a mobile vision sensor as a device able to provide useful information for realizing closed loop control schemes with respect to the environment. Taking into account this particular goal requires:

- the ability to extract from an image the information which is sufficient and pertinent for the completion of the task (in general a positioning task). Note that this may require the design of dedicated targets;
- the implementation of control algorithms simple enough to work at a rate compatible with the desired bandwith of the closed-loop system, but, further, with an acceptable robustness with regard to unavoidable uncertainties existing about the sensor and the environment.

This approach, known as *visual servoing*, is the central issue of this paper.

It should be emphasized that this approach differs from the ones referred to as *Dynamic Vision* approaches, which exploit without controlling it the motion of the sensor or of the objects in the scene. In the *Visual servoing* approach, a first step consists in defining in the image a particular set of characteristic features which constitutes the goal to be reached. A second step will be to design a control which will ensure the convergence towards the configuration corresponding to the goal image, by starting from a different initial condition.

Early studies in this domain, in the 70's, were mainly based on heuristics [1], [17]. More formalized approaches arose around 1982, setting out two kinds of problems:

- the choice and the extraction of the visual feature elements to be used in the control;
- the analysis and the synthesis of the control schemes with the point of view of automatic control theory.

For the first point, it should be noticed that the characteristic parameters allowing to define the goal image were often selected in relation to the existence of algorithms allowing their extraction within a reasonably short time interval. This is why it is not surprising that most of the performed works use low level primitives like *contours*: points, segments [19], [11], [15] or *regions*: surface, barycenter, inertia axis [17], [29]. The original approach of [25] should also be cited: here, the image of a polyhedron is described by a graph, the nodes of which represent the surfaces of the faces and the vertices the lengths of the edges.

For the second point, the most relevant results in the literature come from Carnegie Mellon University [26], [29], [30]. Two kinds of studies have been conducted: the analysis of the mapping between the screw space of the camera motion and the space of velocity fields in the image (which will be called later 'interaction screw'), and the design of control schemes with the study of their stability. Concerning the first point, results about the *sensitivity* with respect to the choice of image features were obtained [30], [13], [18]. Excepted for [13], the results concerned simplified cases: pure translational camera motion, assumption of regularity of the mapping. It should also be noticed that no general methodology for deriving an analytical form of this mapping has been proposed (in [18], for example, the Jacobian matrix of the mapping is estimated using a perturbation method, in [13], only points are studied). In [30], the analysis of control laws is investigated with three points of view: decoupling, adaptive control and model-based control. A stability analysis leading to the synthesis of a controller can also be found in [29] and [9].

It appears that the referenced works do not in general investigate in depth the choice of the visual features to be used in connection with the applications. Furthermore, the related analysis of control schemes is not inserted in an overall approach of the control problem, taking into account specific aspects of robotics, like the effects of modeling errors or approximations.

In this paper, the problem of 'visual servoing' is studied with a more general point of view, under two aspects. A first point will consist in proposing a general methodology for the design of tasks which use a visual sensor inside the control loop. This first modeling stage will be followed by an analysis of control aspects, focused on the study of *robustness* properties of a control scheme. This analysis, partially reported in [20], is based on the original approach of the robot control problem of Samson [24]. The objective of the control is, in the present case, the robust positioning of a mobile camera with regard to the environment, with a task directly expressed as an error with respect to a goal image.

The paper is organized as follows. Section 2 is devoted to modeling questions: after having derived a general model of the interaction between a sensor and a rigid environment, the special case of images is treated, and four relevant examples are examined: points, lines, circles and spheres. In section 3, control aspects are considered: in 3.1, the basics are recalled: task functions, redundancy, hybrid tasks, control schemes, stability issues. Section 3.2 presents the application of this general approach for using a camera. Finally, section 4 gives several experimental and simulation results.

2 Modeling

2.1 Generalities

Notation

The following notation will be employed:

Let us consider the tri-dimensional affine euclidean space, the related vector space being $\mathbb{R}^3$. The configuration space of a rigid body, which is also the frame configuration space, is the Lie group of displacements, SE_3 (Special Euclidean Group). It is a six-dimensional differential manifold. An element of SE_3, called a 'location' (i.e. position and attitude owing to the previous isomorphism) is denoted as $\bar{r}$. The tangent space to SE_3 at identity is denoted as se_3, and its dual, or cotangent space, se_3^*. se_3 is a Lie algebra isomorphic to the Lie algebra of equiprojective fields in $\mathbb{R}^3$ and an element (field) of se_3 is also known as a classical velocity screw. A screw H is also defined by its vector u and the value of its field in a point O, such that, for any P:

$$H(P) = H(O) + u \times OP \qquad (1)$$

We will therefore simply write: $H = (H(O), u)$.

For any considered point O, the screw product is defined by:

$$H_1 \bullet H_2 = <u_1, H_2(O)> + <u_2, H_1(O)> \qquad (2)$$

where $<,>$ is the usual scalar product between two vectors of $\mathbb{R}^3$. Let S be a screw space. The screw product induces an isomorphism between S and its dual S^*. So, S^* is itself a screw space.

The Interaction Screw

Let us now examine the mathematical concepts which may be associated to a sensor. We will indeed restrict our study to the class of sensors such that, formally:

– PROPERTY P_1: *a sensor is completely defined by a differentiable mapping from SE_3 to $\mathbb{R}^p$.*

This property implies in particular that, for a given sensor, relative environmental modifications of the geometrical kind are the only ones allowed to make the sensor output vary. Let us link a frame F_E to the part of environment observed by the sensor, and another, F_S, to the sensor itself. The sensor output s can then be written: $s(F_S, F_E)$. Let us now consider that the sensor motion is obtained by using a 6 degrees of freedom physical device. Its configuration is represented by a generalized coordinate system, q, which is assumed to be a local chart of SE_3 (a submanifold of SE_3 might also be considered). Then, when the observed objects are autonomously mobile themselves, s may be also written $s = s(q, t)$, the independent time variable t representing the contribution of the objects motion. The six variables q_i are for example the joint angular positions of a rigid manipulator which handles a camera.

Let us now examine a component s_j of s. Owing to the above preliminaries and to P_1, we know that its differential at $\bar{r}$, $ds_{j|_{\bar{r}}}$, is a linear mapping from $se_{3|_{\bar{r}}}$ to $\mathbb{R}$. It is also known that the differential of any analytic function from a manifold to $\mathbb{R}$ can be identified with an element of the cotangent space. In our case, this implies that the differential of s_j at $\bar{r}$ is simply an element of se_3^*, that is to say a screw. Recalling that an element of se_3 is the velocity screw, we can therefore write at $\bar{r}$ in SE_3 the basic relation:

$$\dot{s}_j = H_j \bullet T_{SE} \qquad (3)$$

where:

- T_{SE} is the velocity screw of the frame F_E with respect to the frame F_S;
- • is the screw product defined above;
- H_j is a screw, the expression of which depends both on the environment characteristics and on the sensor itself. It therefore fully characterizes the interaction between a sensor and its environment.

We can then set:

- DEFINITION D_1: Under property P_1, the screw H_j defined by the expression $\dot{s}_j = H_j \bullet T_{SE}$ is called the *interaction screw*.

With an obvious breach of notation, equation (3) may also be written:

$$\dot{s}_j = L_j^T \, T_{SE} \quad \text{where} \quad L_j^T = H_j \begin{pmatrix} 0 & \mathbf{I}_3 \\ \mathbf{I}_3 & 0 \end{pmatrix} \tag{4}$$

L_j^T is the matrix-form of the interaction screw H_j, in a given frame F and in a chosen point O. Therefore:

- DEFINITION D_2: the matrix form of the set $\{H_1 \cdots H_p\}$ is called *Interaction Matrix*, and is denoted as L^T.

It will appear later that the interaction matrix plays an important role in modeling aspects as well as in the control itself.

The Concept of Virtual Linkage

Let T^* be a virtual motion at $\bar{r}$ keeping constant the sensor output s_j. T^* is solution of the equation:

$$H_j \bullet T^* = 0 \tag{5}$$

and is therefore a screw *reciprocal* to H_j. The set of the motions T^* leaving the p components of s invariant is S^*, the subspace reciprocal to the screw subspace S spanned by the set $\{H_1 \cdots H_p\}$ (using the interaction matrix, we have $S^* = \text{Ker } L^T$).

More precisely, we can say that:

- DEFINITION D_3: A set of *compatible* and *independent* constraints, $s(\bar{r}) - s_d(t) = 0$, constitutes a *virtual linkage* between the sensor and the objects of the environment.
  Let m be the dimension of S. In a location where these constraints are satisfied, the dimension, $N = 6 - m$, of S^* is called the *class* of the virtual linkage in $\bar{r}$ and at time t.

This concept is an immediate extension of the basic kinematics of contacts, as classically used in the theory of mechanisms. The concept of virtual linkage will allow us later to design the sensor-referenced robotics tasks in a simple way. This also establishes a connection with the approach known as 'hybrid control', which is traditionally used in control schemes involving contact force sensors.

- **Remark:** When $m = p$, the dimension of the signal vector s is adequate, in the intuitive sense that it is indeed the number of 'degrees of freedom' to be controlled from s. However, the case $p > m$ often offers some practical advantages, for example because of the filtering effects it may induce. We will therefore explicitly include this case in the control design.

2.2 Case of an Image Sensor

Framework

Let us reduce a camera to a perspective projection model (figure 1).

[hbtp]

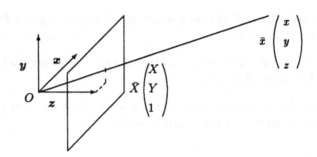

Fig. 1. A simple camera model

The velocity screw of the camera frame $F_S(O, \boldsymbol{x}, \boldsymbol{y}, \boldsymbol{z})$, given figure 1, with respect to the scene is denoted as $T = (v(O), \omega)$ where $v(O)$ and ω are respectively the translational and rotational velocities. In the following, all the used variables (point coordinates, screws,...) will be assumed to be expressed in that frame.

Without loss of generality, the focal length is assumed to be equal to 1, so that any point with coordinates $\bar{x} = (x\,y\,z)^T$ is projected on the image plane as a point with coordinates $\bar{X} = (X\,Y\,1)^T$ with:

$$\bar{X} = \frac{1}{z}\,\bar{x} \qquad (6)$$

- **Remark:** It appears that this geometrical model should be accompanied with a photometric model. However, in this paper, we consider that measurements involving photometric parameters would be inadequate to constitute sensor signals as defined previously. Indeed, property P_1 is generally violated with such measurements, because of their large sensitivity to the lighting parameters which may vary independently of the camera or of the objects motions.

Moreover, the derivation of the interaction screw from a photometric model appears to be hazardous because of drastic uncertainties: the usual simplifying assumptions (Lambertian surfaces, diffuse lighting) are in general unrealistic, and the induced errors would then make the computation of the interaction screw irrelevant.

Basic Modeling Issues

We examine in this paragraph the different components which form the visual features s. In particular, we set the conditions under which the property P_1 is satisfied. Let us firstly state (see figure 2):

- DEFINITION D_4: a *scene feature* is a set of tri-dimensional geometrical primitives (points, lines, vertices...) rigidly linked to a single body. A *configuration* of the scene feature is an element P_S of the n_S-dimensional set $\mathcal{P}_s$ of all possible configurations ($n_S \leq 6$).

[htbp]

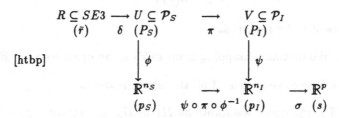

Fig. 2. From relative location to image features

Let π be the perspective projection mapping. Then:

- DEFINITION D_5: an *image feature* is a set of primitives in the image plane which corresponds to the projection of a scene feature. A *configuration* of the image feature is an element P_I, with $P_I = \pi(P_S)$, of the n_I-dimensional set $\mathcal{P}_I$ of all possible configurations.

Furthermore, the group of displacements acts on $\mathcal{P}_S$ through the mapping δ such that:

$$P_S = \delta(\bar{r}) \tag{7}$$

Let us now assume that:

- H_1: displacements are restricted to an open set R of SE_3 such that, $\forall \bar{r} \in R$, none degenerated case (where, for example, a line projects onto the image plane as a point or a circle as a segment) occurs.

Let U be an open set of $\mathcal{P}_S$ including P_S such that $U = \delta(R)$ and let V be an open set of $\mathcal{P}_I$ such that $V = \pi(U)$. We are now ready to state the following assumption:

– H_2: U and V are differential manifolds.

which allows us to define two local charts [27]:

- around P_S: (U, ϕ): $p_S = \phi(P_S)$ where $p_S \in \mathbf{R}^{ns}$ is a parametrization of P_S (for example, if P_S is a point, p_S may be chosen as its three coordinates);
- around P_I: (V, ψ): $p_I = \psi(P_I)$ where $p_I \in \mathbf{R}^{ni}$ is a parametrization of P_I (with the same example, p_I may be chosen as the two coordinates in the image plane of the projection of the point).

Therefore:

$$p_I(p_S) = \psi \circ \pi \circ \phi^{-1}(p_S) \tag{8}$$

and:

$$p_I(\bar{r}) = \psi \circ \pi \circ \delta(\bar{r}) \tag{9}$$

Now, s has to be chosen such that p_I is able to generate an useful sensor signal. The associated mapping is:

$$s = \sigma(p_I) \tag{10}$$

which is supposed to be such that:

– H_3: σ is a differentiable mapping from $\psi(V)$ to an open set Ω of $\mathbf{R}^p$.

Now, an immediate consequence of all the above issues is:

– PROPERTY P_2: under assumptions H_1 to H_3, the sensor signal s defined in (10) belongs to the class specified by property P_1 on the open set Ω of $\mathbf{R}^p$. Furthermore, we have formally:

$$\frac{\partial s}{\partial \bar{r}} = \frac{\partial \sigma}{\partial p_I} \frac{\partial \psi}{\partial P_I} \frac{\partial \pi}{\partial P_S} \frac{\partial \delta}{\partial \bar{r}} \tag{11}$$

This form is not the most adequate to an analytic computation of the interaction matrix and the method described in the next paragraph is often preferred.

Derivation of the interaction matrix

Let us specify the configuration $\mathcal{P}_{S_j}$ of the j-th primitive of a scene feature by an equation of type:

$$h(\bar{x}, \bar{p}) = 0, \quad \forall \bar{x} \in \mathcal{P}_{S_j} \tag{12}$$

where h defines the kind of the primitive and the value of $\bar{p}$ corresponds to its configuration, and where $\forall \bar{x} \in \mathcal{P}_{S_j}$ means 'for any point of the tri-dimensional space, with coordinates $\bar{x}$, belonging to the geometrical primitive'.

– **Remark:** If the scene feature is constituted by only one primitive and if $\bar{p}$ is chosen as a chart-representation (which is not here required), we have $\bar{p} = p_S$.

By using (6), equation (12) becomes:

$$h(z\bar{X}, \bar{p}) = 0 \qquad (13)$$

i.e.

$$h'(\bar{X}, z, \bar{p}) = 0 \qquad (14)$$

Under the trivial condition $\frac{\partial h'}{\partial z} \neq 0$ which is satisfied in all the non-degenerated cases, the implicit function theorem ensures the existence of a unique function μ around a solution $\bar{x}_0$ of (12), such that:

$$z = \mu(\bar{X}, \bar{p}) \qquad (15)$$

Using (15) in (14) allows to write the configuration $\mathcal{P}_{I_j}$ corresponding to the j-th primitive of the image feature under the form:

$$g(\bar{X}, \bar{P}) = 0, \quad \forall \bar{X} \in \mathcal{P}_{I_j} \qquad (16)$$

where the set of parameters $\bar{P}$, obtained from $\bar{p}$, is of dimension l and where $\forall \bar{X} \in \mathcal{P}_{I_j}$ means 'for any point of the image plane, with coordinates $\bar{X}$, belonging to the geometrical image primitive'.

In practice, the sensor s is chosen as a differentiable function of $\bar{P}$: $s = f(\bar{P})$. Therefore:

$$\frac{\partial s}{\partial \bar{r}} = \frac{\partial s}{\partial \bar{P}} \frac{\partial \bar{P}}{\partial \bar{p}} \frac{\partial \bar{p}}{\partial \bar{r}} \qquad (17)$$

The computation of $\frac{\partial s}{\partial \bar{P}}$ and $\frac{\partial \bar{P}}{\partial \bar{p}}$ is generally trivial. On the other hand, the term $\frac{\partial \bar{p}}{\partial \bar{r}}$ is not always computable. In that case, we have to use another method to directly compute $\frac{\partial \bar{P}}{\partial \bar{r}}$.

- **Remark:** When s depends on the configuration of k differents primitives (as, for example, distance between two points or orientation between two lines), we have $s = f(\bar{P}_1, \cdots, \bar{P}_k)$ and $\frac{\partial s}{\partial \bar{r}}$ can be obtained in a similar way:

$$\frac{\partial s}{\partial \bar{r}} = \sum_{i=1}^{k} \frac{\partial s}{\partial \bar{P}_i} \frac{\partial \bar{P}_i}{\partial \bar{r}} \qquad (18)$$

Knowing that the rigidity assumption implies:

$$\dot{g} = 0, \quad \forall \bar{X} \in \mathcal{P}_{I_j} \qquad (19)$$

we can now compute the interaction matrix L^T associated to $\bar{P}$. Differentiation of (16) gives:

$$\frac{\partial g}{\partial \bar{P}}(\bar{X}, \bar{P}) \, \dot{\bar{P}} = -\frac{\partial g}{\partial \bar{X}}(\bar{X}, \bar{P}) \, \dot{\bar{X}}, \quad \forall \bar{X} \in \mathcal{P}_{I_j} \qquad (20)$$

Differentiating (6) leads to the well known optical flow equations, which can be written:

$$\dot{\bar{X}} = L_{of}^T(\bar{X}, z) \, T \qquad (21)$$

115

where T is, as previously set, the velocity of the camera with respect to the scene, and with:

$$L_{of}^T = \begin{pmatrix} -1/z & 0 & X/z & XY & -(1+X^2) & Y \\ 0 & -1/z & Y/z & 1+Y^2 & -XY & -X \end{pmatrix} \tag{22}$$

Using (15) in (22), gives:

$$L_{of}^T(\bar{X}, z) = L_{of}'^T(\bar{X}, \bar{p}) \tag{23}$$

Finally, (20), (21) and (23) lead to:

$$\frac{\partial g}{\partial \bar{P}}(\bar{X}, \bar{P})\,\dot{\bar{P}} = -\frac{\partial g}{\partial \bar{X}}(\bar{X}, \bar{P})\,L_{of}'^T(\bar{X}, \bar{p})\,T\ ,\ \forall \bar{X} \in \mathcal{P}_{I_j} \tag{24}$$

This equation can be solved, either by explicitly using (16) in (24) and identifying, or in the following way:

The matrix L^T is of dimension $l \times 6$. By choosing l points with coordinates $\bar{X}_i$ belonging to the image primitive, we get:

$$\begin{cases} \dfrac{\partial g}{\partial \bar{P}}(\bar{X}_i, \bar{P}) = \alpha_i^T(\bar{P}) \\[2mm] -\dfrac{\partial g}{\partial \bar{X}}(\bar{X}_i, \bar{P}) L_{of}'^T(\bar{X}_i, \bar{p}) = \beta_i^T(\bar{p}, \bar{P}) \end{cases} \tag{25}$$

The interaction matrix is then:

$$L^T(\bar{p}, \bar{P}) = \begin{pmatrix} \alpha_1^T \\ \vdots \\ \alpha_l^T \end{pmatrix}^{-1} (\bar{P}) \begin{pmatrix} \beta_1^T \\ \vdots \\ \beta_l^T \end{pmatrix} (\bar{p}, \bar{P}) \tag{26}$$

- **Remark:** It should be emphasized that the computation of L^T using (26) requires the non-singularity of the matrix $(\alpha_1 \cdots \alpha_l)$. This is ensured when the dimension l of the parametrization $\bar{P}$ is minimal (i.e. the l chosen points belonging to the image primitive are independent).

Examples of usual features

Let us now examine a few basic primitives:

Points. Consider a point m_i with coordinates $\bar{x}_i$. Then, $\bar{p} = \bar{x}_i = (x_i\,y_i\,z_i)^T$ and $\bar{P} = \bar{X}_i = (X_i\,Y_i)^T$ where X_i and Y_i are the coordinates of the projection of m_i in the image. The interaction matrices L_{X_i} and L_{Y_i} related to X_i and Y_i are given by (22) and are written:

$$\begin{matrix} L_{X_i}^T = [\,-1/z_i & 0 & X_i/z_i & X_iY_i & -(1+X_i^2) & Y_i\,] \\ L_{Y_i}^T = [\ 0 & -1/z_i & Y_i/z_i & 1+Y_i^2 & -X_iY_i & -X_i\,] \end{matrix} \tag{27}$$

Various sensor signals can be generated from image points [6]: for example, length and orientation of a segment, surface and mass center of a polygon, etc .

Straight lines. A straight line can be represented as the intersection of two planes:

$$h(\bar{x}, \bar{p}) = \begin{cases} a_1 x + b_1 y + c_1 z + d_1 = 0 \\ a_2 x + b_2 y + c_2 z + d_2 = 0 \end{cases} \tag{28}$$

If we exclude the degenerated case where the projection center belongs to the straight line $(d_1 = d_2 = 0)$, the equation of the projected line in the image plane is:

$$AX + BY + C = 0 \text{ with } \begin{cases} A = a_1 d_2 - a_2 d_1 \\ B = b_1 d_2 - b_2 d_1 \\ C = c_1 d_2 - c_2 d_1 \end{cases} \tag{29}$$

Since the parametrization (A, B, C) of 2D straight lines is not minimal, another one must be preferred. The most used, $\bar{P} = (a, b)$, is inadequate because two non compatible charts $(Y = aX + b, \ X = aY + b)$ have to be used. We therefore choose $\bar{P} = (\rho, \theta)$ and the equation of a straight line $\mathcal{D}$ is then:

$$g(\bar{X}, \bar{P}) = X \cos \theta + Y \sin \theta - \rho = 0 \tag{30}$$

- **Remark:** The ambiguity of this representation (the same line may be parametrized by $(\rho, \theta + 2k\pi)$ and $(-\rho, \theta + (2k + 1)\pi)$) is partly overcome by defining a direction for the lines, i.e fixing the sign of ρ. It will be seen that the ambiguity on the multiple possible choices of θ has no consequences on the values of the interaction screws. It can also be easily overcome as shown in section 4.2.2.

Let us now derive the interaction screws related to this parametrization. Equation (20) can then be written:

$$\dot{\rho} + (X \sin \theta - Y \cos \theta)\dot{\theta} = \cos \theta \dot{X} + \sin \theta \dot{Y} \ , \ \forall (X, Y) \in \mathcal{D} \tag{31}$$

Furthermore, the function μ of (15) is obtained from h_1 or h_2:

$$1/z = -(a_i X + b_i Y + c_i)/d_i \text{ with } i = 1 \text{ if } d_1 \neq 0 \text{ or } i = 2 \text{ if } d_2 \neq 0 \tag{32}$$

From (30), X is expressed as a function of Y when $\cos \theta \neq 0$ (or Y as a function of X in the other case). By using (22) and (32), equation (31) is written:

$$(-\dot{\theta}/\cos \theta)\, Y + (\dot{\rho} + \rho \tan \theta \ \dot{\theta}) = Y \ K_1 T + K_2 T \ , \ \forall Y \in \mathbb{R} \tag{33}$$

where

$$\begin{cases} K_1 = [\lambda_1 \cos \theta \ \lambda_1 \sin \theta \ -\lambda_1 \rho \quad \rho \qquad \rho \tan \theta \qquad 1/\cos \theta] \\ K_2 = [\lambda_2 \cos \theta \ \lambda_2 \sin \theta \ -\lambda_2 \rho \sin \theta - \cos \theta - \rho^2/\cos \theta -\rho \tan \theta] \end{cases}$$

and $\lambda_1 = (-a_i \tan \theta + b_i)/d_i, \ \lambda_2 = (a_i \rho/\cos \theta + c_i)/d_i$.

This leads to:

$$\begin{cases} \dot{\theta} = -K_1 \ \cos \theta \ T \\ \dot{\rho} = (K_2 + \rho \sin \theta K_1) \ T \end{cases} \tag{34}$$

We have therefore:

$$\begin{aligned} L_\theta^T &= [\ \lambda_\theta \cos \theta \ \lambda_\theta \sin \theta \ -\lambda_\theta \rho \quad -\rho \cos \theta \qquad -\rho \sin \theta \quad -1\ ] \\ L_\rho^T &= [\ \lambda_\rho \cos \theta \ \lambda_\rho \sin \theta \ -\lambda_\rho \rho \ (1 + \rho^2) \sin \theta \ -(1 + \rho^2) \cos \theta \ \ 0\ ] \end{aligned} \tag{35}$$

with $\lambda_\theta = (a_i \sin \theta - b_i \cos \theta)/d_i$ and $\lambda_\rho = (a_i \rho \cos \theta + b_i \rho \sin \theta + c_i)/d_i$.

Plane primitives. h is then two-dimensional ($h = (h_1\ h_2)^T$). Let us for example choose the equation of the plane in which the primitive lies for h_2. The function μ of (15) is then obtained from $h'_2(\bar{X}, z, \bar{p}) = 0$. With (15), $h'_1(\bar{X}, z; \bar{p}) = 0$ gives:

$$\tilde{h}(\bar{X}, \bar{p}) = 0 \quad \text{with dim } \tilde{h} = 1 \qquad (36)$$

which, after change of parametrization, is written $g(\bar{X}, \bar{q}) = 0$. Let us now consider an example of plane primitive: the circle.

Circles. A circle may be represented as the intersection of a sphere and a plane:

$$h(\bar{x}, \bar{p}) = \begin{cases} (x - x_0)^2 + (y - y_0)^2 + (z - z_0)^2 - r^2 = 0 \\ \alpha(x - x_0) + \beta(y - y_0) + \gamma(z - z_0) = 0 \end{cases} \qquad (37)$$

Its projection on the image plane takes the form of an ellipse (except in degenerated cases where the projection is a segment), which may be represented in a non-ambiguous form as:

$$g(\bar{X}, \bar{P}) : X^2 + A_1 Y^2 + 2A_2 XY + 2A_3 X + 2A_4 Y + A_5 = 0 \qquad (38)$$

$$\text{with} \begin{cases} A_1 = [b^2(x_0^2 + y_0^2 + z_0^2 - r^2) + 1 - 2by_0]/A_0 \\ A_2 = [ab(x_0^2 + y_0^2 + z_0^2 - r^2) - bx_0 - ay_0]/A_0 \\ A_3 = [ac(x_0^2 + y_0^2 + z_0^2 - r^2) - cx_0 - az_0]/A_0 \\ A_4 = [bc(x_0^2 + y_0^2 + z_0^2 - r^2) - cy_0 - bz_0]/A_0 \\ A_5 = [c^2(x_0^2 + y_0^2 + z_0^2 - r^2) + 1 - 2cz_0]/A_0 \end{cases}$$

where $A_0 = a^2(x_0^2 + y_0^2 + z_0^2 - r^2) + 1 - 2ax_0 \quad (\neq 0)$
and $a = \alpha/(\alpha x_0 + \beta y_0 + \gamma z_0)$, $b = \beta/(\alpha x_0 + \beta y_0 + \gamma z_0)$, $c = \gamma/(\alpha x_0 + \beta y_0 + \gamma z_0)$

The function $\mu(\bar{X}, \bar{p})$, required for the interaction screw derivation, is easily obtained from h_2:

$$1/z = aX + bY + c \qquad (39)$$

then, we can obtain (see [6] for more details):

$$L_{A_1}^T = [\ \ 2bA_2 - 2aA_1 \quad\quad 2A_1(b - aA_2) \quad\quad 2bA_4 - 2aA_1A_3$$
$$2A_4 \quad\quad\quad 2A_1A_3 \quad\quad\quad -2A_2(A_1 + 1) \quad\quad]$$

$$L_{A_2}^T = [\ \ b - aA_2 \quad\quad bA_2 - a(2A_2^2 - A_1) \quad a(A_4 - 2A_2A_3) + bA_3$$
$$A_3 \quad\quad\quad 2A_2A_3 - A_4 \quad\quad\quad A_1 - 2A_2^2 - 1 \quad\quad]$$

$$L_{A_3}^T = [\ \ c - aA_3 \quad\quad a(A_4 - 2A_2A_3) + cA_2 \quad cA_3 - a(2A_3^2 - A_5)$$
$$-A_2 \quad\quad\quad 1 + 2A_3^2 - A_5 \quad\quad\quad A_4 - 2A_2A_3 \quad\quad]$$

$$L_{A_4}^T = [\ A_3 b + A_2 c - 2aA_4 \ \ A_4 b + A_1 c - 2aA_2A_4 \ \ bA_5 + cA_4 - 2aA_3A_4$$
$$A_5 - A_1 \quad\quad\quad 2A_3A_4 + A_2 \quad\quad\quad -2A_2A_4 - A_3 \quad\quad]$$

$$L_{A_5}^T = [\ \ 2cA_3 - 2aA_5 \quad\quad 2cA_4 - 2aA_2A_5 \quad\quad 2cA_5 - 2aA_3A_5$$
$$-2A_4 \quad\quad\quad 2A_3A_5 + 2A_3 \quad\quad\quad -2A_2A_5 \quad\quad]$$

$$(40)$$

– **Remark:** The interaction matrix L^T related to an ellipse is always of rank 5 excepted when the projection of a circle is a *centered* circle in the image plane. In that case, we have:

$$A_1 = 1, A_2 = A_3 = A_4 = 0, A_5 = -r^2/z_0^2 \Leftrightarrow a = b = x_0 = y_0 = 0 \quad (41)$$

L^T is here of rank 3. This singularity (loss of rank) is isolated, since the rank of the matrix L^T remains 5, even in the other cases when the projection is a circle:

$$A_1 = 1, A_2 = 0 \Leftrightarrow \begin{cases} a = b = 0 \\ \text{or} \\ a = 2x_0/(x_0^2 + y_0^2 + z_0^2 - r^2)\,, \\ b = 2y_0/(x_0^2 + y_0^2 + z_0^2 - r^2) \end{cases} \quad (42)$$

Tri-dimensional primitives. Let us now consider spheres, cylinders, torus,... (or even intersection of some tri-dimensional primitives). In that case, it is often possible to exhibit the functions μ and g given by (15) and (16). We then consider the contour of the primitive projected on the image. The function $g(\bar{X}, \bar{P})$ is thus the limb equation, and the matching between 3D points and contour points may be expressed as $h_0'(\bar{X}, z, \bar{p}) = 0$, with dim $h_0' = 1$, which provides function $z = \mu(\bar{X}, \bar{p})$.

Spheres. A sphere is an example of tri-dimensional primitive, with equation:

$$h(\bar{x}, \bar{p}) = (x - x_0)^2 + (y - y_0)^2 + (z - z_0)^2 - r^2 = 0 \quad (43)$$

The functions $g(\bar{X}, \bar{P})$ and $\mu(\bar{X}, \bar{p})$ have to be determined in a first step. By using (6) in (43), we obtain a polynomial having $1/z$ as variable:

$$\frac{1}{z^2}(x_0^2 + y_0^2 + z_0^2 - R^2) - \frac{2}{z}(x_0 X + y_0 Y + z_0) + X^2 + Y^2 + 1 = 0 \quad (44)$$

Points belonging to the image contour of a sphere are such that the intersection between their viewline and the sphere is unique. In the present case, it is equivalent to say that the discriminant Δ of (44) vanishes. Therefore, we have:

$$(x_0 X + y_0 Y + z_0)^2 - (x_0^2 + y_0^2 + z_0^2 - R^2)(X^2 + Y^2 + 1) = 0 \quad (45)$$

which can be written after a change of parametrization:

$$g(\bar{X}, \bar{P}): X^2 + A_1 Y^2 + 2A_2 XY + 2A_3 X + 2A_4 Y + A_5 = 0 \quad (46)$$

$$\text{with} \begin{cases} A_1 = (R^2 - x_0^2 - z_0^2)/(R^2 - y_0^2 - z_0^2) \\ A_2 = x_0 y_0/(R^2 - y_0^2 - z_0^2) \\ A_3 = x_0 z_0/(R^2 - y_0^2 - z_0^2) \\ A_4 = y_0 z_0/(R^2 - y_0^2 - z_0^2) \\ A_5 = (R^2 - x_0^2 - y_0^2)/(R^2 - y_0^2 - z_0^2) \end{cases}$$

The image of a sphere is thus an ellipse (a circle when $x_0 = y_0 = 0$).

The μ function is easily obtained from the double root of (44) corresponding to $\Delta = 0$ and we have:

$$1/z = aX + bY + c \text{ with } \begin{cases} a = x_0/(x_0^2 + y_0^2 + z_0^2 - R^2) \\ b = y_0/(x_0^2 + y_0^2 + z_0^2 - R^2) \\ c = z_0/(x_0^2 + y_0^2 + z_0^2 - R^2) \end{cases} \qquad (47)$$

The derivation of interaction screws related to an ellipse represented by $\bar{P} = (A_1, ..., A_5)^T$ was already done for the circle. Results are therefore given by equations (40) with the particular above values for a, b and c. The related matrix L^T is always of rank 3 for any sphere configuration. Since three parameters are sufficient to characterize the image of a sphere, we may choose the set (A_3, A_4, A_5) because the associated interaction matrix is then always of full rank, contrary to the use of A_1 or A_2.

Similar results may be obtain for others 3D primitives. For example, cylinders are also studied in [6].

3 Control

3.1 A General Framework

This work is embedded in the larger domain of sensor-based control. A general approach of the problem is presented by C. Samson in [21] and [24] and, since all developments may be found in these references, we will only recall here its main characteristics, without any proof.

The concept of task function

It is known that the dynamic behavior of a rigid manipulator is described by the model:

$$\Gamma = M(q)\ddot{q} + N(q, \dot{q}, t), \quad \dim(q) = \dim(M) = n \qquad (48)$$

where:

- Γ is the vector of applied external forces (actuator torques);
- M is the kinetics energy matrix;
- N gathers gravity, centrifugal, Coriolis and friction forces.

Equation (48) is the *state* equation of the system, with the natural associated state $(q, \dot{q})$ (it is assumed that an actuator is associated to every degree of freedom of the robot). The task to be performed may then be specified as an *output function* associated to (48). The problem is indeed well-conditioned if the passage between the 'control space' and the 'output space' is regular in some sense.

More precisely, we can show [24] that the user's objective may in general be expressed as the regulation to zero of some n-dimensional C^2 function, $e(q, t)$, called *task function*, during a time interval $[0, t_m]$. Several cases of task functions are presented in [24]. When sensors are used, it appears that the sensor vector $s(q, t)$ has to contribute to the design of $e(q, t)$, in a way explained later.

The problem of regulating e is well-conditioned if e presents some specific properties. One of them is the existence and the uniqueness of a C^2 *ideal trajectory*, $q_r(t)$, such that $e(q_r(t), t) = 0$, $\forall t \in [0, t_m]$ and $q_r(0) = q_0$, where q_0 is a given initial condition. Another very important one, is the non-singularity of the task-Jacobian matrix $\frac{\partial e}{\partial q}(q, t)$, around $q_r(t)$. When all the required conditions are satisfied, the task function is said to be 'admissible', which then allows the realization of efficient control laws more or less robust according to the task admissibility.

Control and stability

We only give here an intuitive idea of the used approach and of the obtained results. Let us consider the problem of exact decoupling and feedback linearization in the task space. By differentiating twice $e(q, t)$, we obtain:

$$\ddot{e}(q, t) = \frac{\partial e}{\partial q}(q, t) \, \ddot{q} + l(q, \dot{q}, t) \tag{49}$$

with:

$$l(q, \dot{q}, t) = \left[\begin{array}{c} \vdots \\ \dot{q}^T \frac{\partial E_i^T}{\partial q}(q, t) \, \dot{q} \\ \vdots \end{array} \right] + 2 \frac{\partial^2 e}{\partial q \partial t}(q, t) \, \dot{q} + \frac{\partial^2 e}{\partial^2 t}(q, t) \tag{50}$$

where $E_i (i = 1, \cdots, n)$ is the i-th row of $\frac{\partial e}{\partial q}$.

Since e is assumed to be admissible, equations (48) and (49) may be combined, which leads to:

$$\Gamma = M \left(\frac{\partial e}{\partial q} \right)^{-1} \ddot{e} + N - M \left(\frac{\partial e}{\partial q} \right)^{-1} l \tag{51}$$

A control which decouples and linearizes in the task space is:

$$\Gamma = M \left(\frac{\partial e}{\partial q} \right)^{-1} u + N - M \left(\frac{\partial e}{\partial q} \right)^{-1} l \tag{52}$$

where u is for example a proportional-derivative feedback of the form:

$$u = -k \, G \, (\mu \, D \, e + \dot{e}) \tag{53}$$

G and D being positive matrices, k and μ being positive scalars, all to be tuned by the user.

The ideal control scheme (52) (53) requires a perfect knowledge of all its components, which is neither possible, nor even wished. A more realistic approach consists in generalizing the previous control as:

$$\Gamma = -k\,\hat{M}\left(\widehat{\frac{\partial e}{\partial q}}\right)^{-1}G\left(\mu\,D\,e + \widehat{\frac{\partial e}{\partial q}}\,\dot{q} + \widehat{\frac{\partial e}{\partial t}}\right) + \hat{N} - \hat{M}\left(\widehat{\frac{\partial e}{\partial q}}\right)^{-1}\hat{l} \qquad (54)$$

where the carets point out that models (approximations, estimates) are used instead of the true terms.

In this general expression, all the terms but μ, D and G are allowed to be functions of q and t, even of $\dot{q}$ for $k, \hat{l}$ and $\hat{N}$. The control (54) includes most of existing schemes: computed torque, resolved motion rate or acceleration control, indirect adaptive control,...

An original theorem concerning the *global* stability of the system (48) with control (54) was established by Samson [24] in a nonlinear framework. Two main classes of sufficient stability conditions (in the sense of the boundedness of $\|e(t)\|$) were then exhibited:

- **gain** conditions. These tuning parameters leave more or less possibilities to the user. We do not consider them here;
- **modeling** conditions. Among them, those related to the robot dynamics are not too strong in practice, owing to the symmetric-positive definiteness of the kinetics energy matrix. Another one, much more critical and less known, is related to the task, and may be expressed as:

$$\frac{\partial e}{\partial q}\left(\widehat{\frac{\partial e}{\partial q}}\right)^{-1} > 0 \qquad (55)$$

along the ideal trajectory (recall that a matrix A $(n \times n)$ is positive if $x^T A x > 0$, $\forall x \neq 0 \in \mathbb{R}^n$). We will return later to the application of this essential condition, which allows to characterize the robustness of the task itself with regard to uncertainties and approximations.

It can already be noticed that, when we are interested in the motion of the end effector, we can write $\frac{\partial e}{\partial q} = \frac{\partial e}{\partial \bar{r}}\frac{\partial \bar{r}}{\partial q}$, where $\frac{\partial \bar{r}}{\partial q}$ is the classical robot Jacobian matrix. When it is known and nonsingular, as we shall assume afterwards, the choice $\widehat{\frac{\partial e}{\partial q}} = \widehat{\frac{\partial e}{\partial \bar{r}}}\frac{\partial \bar{r}}{\partial q}$ allows condition (55) to be reduced to:

$$\frac{\partial e}{\partial \bar{r}}\left(\widehat{\frac{\partial e}{\partial \bar{r}}}\right)^{-1} > 0 \qquad (56)$$

Hybrid Tasks

Introduction. Regulating sensor signals is generally not the unique user's objective; very often, this task has to be combined with another task such as a trajectory tracking.

Generally, the problem specification leads firstly to defining a sensor-based task vector, $e_1(q,t)$, with $m \leq n$ independent components, the regulation of which constitutes the part of the global task which requires the use of extero-ceptive sensors. How to derive such a vector when using visual sensors will be described later. A second objective, for example a desired sensor motion, might be represented at the first glance by a second $(n-m)$-dimensional vector $e_2(q,t)$.

However, e_1 and e_2 would be gathered in a single task vector $e^T = (e_1^T \ e_2^T)$ having the required properties for being admissible, in particular the non-singularity of $\frac{\partial e}{\partial q}$ along $q_r(t)$. This implies that e_1 and e_2 would be compatible and independent, which intuitively means, in terms of virtual linkage, that the secondary goal $(e_2 = 0)$ could be reached using all and only all the realizable motions left available by the virtual linkage associated to the sensor-based task.

It can indeed be shown that a more efficient way of setting the problem con-sists in embedding it in the framework of *task redundancy*. In this approach, e_1 is considered as priority, and e_2 is defined as the representation of the constrained minimization of a secondary cost function. Let us now recall some basic results in this domain, taken from [22] and [24].

The Redundancy Framework. Let us choose SE_3 (with generic element $\bar{r}$) as configuration space. We assume that $\frac{\partial \bar{r}}{\partial q}$ is nonsingular everywhere needed. Let e_1 be a m-dimensional main task, with Jacobian matrix $J_1 = \frac{\partial e_1}{\partial \bar{r}}$, and let h_s be a secondary cost function to be minimized (the choice of h_s is discussed in [24]).

Minimizing h_s under the constraint $e_1 = 0$ requires the subspace of motions left free by this constraint to be determined. This comes back to knowing the null space of J_1, Ker J_1 (or the range of J_1^T, $R(J_1^T)$) along the ideal trajectory. In other words, it has to be found any $m \times n$ full rank matrix W such that:

$$R(W^T) = R(J_1^T) \tag{57}$$

along the robot's ideal trajectory, $q_r(t)$.

Once this matrix is determined, it is rather easy to show [22], [24] that a task function minimizing h_s under the constraint $e_1 = 0$ is:

$$e = W^+ e_1 + \beta \left(\mathbb{I}_6 - W^+ W \right) \frac{\partial h_s}{\partial \bar{r}}^T \tag{58}$$

where:

- β is a positive scalar;
- W^+ is the pseudo-inverse of W;
- $(\mathbb{I}_6 - W^+ W)$ is an orthogonal projection operator on the null space of W, i.e. on that of J_1.

It clearly appears that the computation of the Jacobian matrix related to (58) possibly required in the control scheme, may be complex. The positivity condi-tion (56) is then of some interest. It can indeed be shown that if, in addition to (57), W satisfies the property:

$$J_1 W^T > 0 \tag{59}$$

along $q_r(t)$, then, under 'normal circumstances' (see [24]) the Jacobian matrix of e in SE_3 is such that:

$$\frac{\partial e}{\partial \bar{r}} \left(\mathbf{I}_6 + \gamma \left(\mathbf{I}_6 - W^+ W \right) \right) > 0 \tag{60}$$

along $q_r(t)$, and $\forall \gamma \geq \gamma_m(\beta) \geq 0$.

The condition (56) is therefore satisfied by taking:

$$\left(\widehat{\frac{\partial e}{\partial \bar{r}}} \right)^{-1} = \left(\mathbf{I}_6 + \gamma \left(\mathbf{I}_6 - W^+ W \right) \right) \tag{61}$$

More, when β is 'small enough', then $\gamma_m = 0$, $\frac{\partial e}{\partial \bar{r}}$ is positive, and we may choose:

$$\widehat{\frac{\partial e}{\partial \bar{r}}} = \mathbf{I}_6 \tag{62}$$

 - **Remark:** When e_1 is made from sensor signals and h_s expresses a trajectory tracking task in SE_3, the task represented by (58) is then called 'hybrid task'. The positivity property, which allows the choice (62) to be made, explains why the classical scheme known as 'hybrid control' may work even if an explicit expression of the sensor signal variation is not used in the control equation.

The Specific Case of Sensor Signals. Let us apply this approach to the use of sensor signals. We are interested in regulating s around a desired value or trajectory $s_d(t)$. Let us recall that the vector s is of dimension p, and the Jacobian of s in SE_3 corresponds to the interaction matrix L^T. The dimension of L is $6 \times p$ and its rank m for $s = s_d(t)$, $N = 6 - m$ being the class of the associated virtual linkage (cf section 2.1.3).

Let $C(t)$ be a 'combination matrix', with dimension $m \times p$, such that the matrix CL^T is of full rank m along $q_r(t)$. The main task can then be written [23]:

$$e_1 = C(t) \left(s(\bar{r}, t) - s_d(t) \right) \tag{63}$$

One of the advantages of the existence of C is the possibility of taking into account more sensors (p) than the actual dimension of the constraints they specify (m).

The Jacobian matrix of e_1 in SE_3 is:

$$J_1 = \frac{\partial e_1}{\partial \bar{r}} = CL^T \tag{64}$$

and we can easily show that $R(J_1^T) = R(L)$ along $q_r(t)$.

Owing to (58), the task to be regulated is finally written:

$$e = W^+ C \left(s(\bar{r}, t) - s_d(t) \right) + \beta \left(\mathbf{I}_6 - W^+ W \right) \frac{\partial h_s}{\partial \bar{r}}^T \tag{65}$$

W must satisfy property (57), which then becomes $R\left(W^T\right) = R(L)$. This also means that the rows of W are made from basis vectors of S (evaluated in the camera frame). Interesting is the case where the subspace S, defined in section 2.1.3, is invariant in all the camera locations for which the virtual linkage constraint $s(\bar{r}, t) - s_d(t) = 0$ is satisfied. It occurs when $s_d(t)$ is constant, $\forall t$, or, more generally, when S^* (the subspace reciprocal to S which is the set of motions keeping constant $s(\bar{r}, t)$) is invariant. In that case, the knowledge of S allows to find W constant which satisfies (57). In other cases, for example when a task consists in a sequence of several virtual linkages, W has only to be evaluated at each change of S.

Finally, property (59) becomes:

$$CL^T W^T > 0 \tag{66}$$

which allows to choose C knowing L and W. It can be shown that condition (66) is for example satisfied by selecting $C = WL$ or $C = WL^{T^+}$. Let us note that this last choice ensures a best behavior of the control law. Indeed, if we simply consider the case $m = 6$, we obtain $\frac{\partial e}{\partial \bar{r}} = \mathbf{I}_6$, instead of LL^T. The choice (62) is thus perfectly justified by setting $C = WL^{T^+}$.

3.2 Application to Visual Sensors

It has been seen that, when data provided by a mobile camera are used as sensor signals, the associated interaction matrices L^T depend both on parameters measured in the image, $\bar{P}$, and on 3D information coming from the considered primitives, $\bar{p}$. Since this last information is a-priori unknown, it is necessary to choose a model $\hat{L}$ of L. Properties (57) and (59) will thus be satisfied if we may ensure that, respectively, $R\left(\hat{L}\right) = R(L)$ and $CL^T W^T > 0$. Recall that the interest of satisfying these properties lies in the possibility of simply choosing the identity matrix as a model of $\frac{\partial e}{\partial \bar{r}}$.

Several possibilities exist:

- $\hat{L} = L(\hat{\bar{p}}, \bar{P})$ when $\bar{p}$ is concurrently estimated.
- $\hat{L} = L(\bar{p}_d, \bar{P})$ where $\bar{p}_d$ is the value of $\bar{p}$ at $s = s_d(t)$ which realizes $e_1 = 0$. Some further assumptions are then needed. We will come back to this point later.
- $\hat{L} = L(\hat{\bar{p}}_d, \bar{P})$ where $\hat{\bar{p}}_d$ is an estimate of $\bar{p}_d$ when no 3D information is available.

With the above choices, the matrix C needs to be updated at the same rate as the control loop. This may be not efficient, for example when C is chosen equal to

$W\hat{L}^{T^+}$, because of the computing time required by the computation of pseudo-inverses. It is also sometimes necessary to anticipate the possible crossing of an isolated singularity (case of the circle in section 2.2.4). Such a crossing would lead to a loss of rank for L^T, thus to a reduction of the dimension of e_1. Taking into account these unexpected singularities might be complex, and a simpler solution consists in using a model $\hat{L}$, determined within the task design step. It may then be chosen:

- $\hat{L} = L(\bar{p}_d, \bar{P}_d)$, also denoted as $L_{|s=s_d}$, which is the value of the interaction matrix at a location corresponding to the selected feature $s = s_d(t)$.

The positivity condition (66) is then often only satisfied in the neighborhood of $s = s_d(t)$, whichever the choice of C. Fortunately, it should be emphasized that this condition is only *sufficient*, and we shall see in the examples of section 4 that the convergence of the control law may be obtained even from initial conditions far away from the goal configuration.

This choice of $\hat{L}$ requires the knowledge of $\bar{p}_d$, which is equivalent to making assumptions about the *shape* and the *geometry* of the 3D scene. Such assumptions may often be done within the task design step, and seem then not too strong: for example, if the task consists in positioning the camera in front of a door, it may be assumed that there is a door in the scene, and that characteristic signals of this door (for example its four corners) may be extracted. In addition, if it is desired to place the camera at a given range of the door, its dimensions should be known in order to determine the goal feature in the image.

Finally, when no scene knowledge is available (for example when it is wished to track an unknown object with a goal image feature extracted from an initial image of the object), it may be chosen:

- $\hat{L} = L(\hat{\bar{p}}_d, \bar{P}_d)$ where $\hat{\bar{p}}_d$ is an estimate of $\bar{p}_d$, not necessarily very accurate, but ensuring the asymptotic stability of the controlled system in some neighborhood of $s = s_d(t)$. Ensuring the positivity condition to be satisfied is then difficult, even when $s = s_d(t)$, since the value $L_{|s=s_d}$ remains unknown.

4 Experimental Results

In the presented cases, the camera is assumed to have six degrees of freedom. The goal consists in positioning the camera with respect to a target. Let us note that shape and geometrical assumptions on the different targets have been made. Therefore, $\hat{L} = L_{|s=s_d}$. Furthermore, we have only considered the case where the vision-based task consists in regulating $s(\bar{r}, t)$ around a desired value s_d, and not a trajectory $s_d(t)$. This limitation allows to choose constant matrices W and C in (65).

Before showing the obtained results, we describe the simplified version of the control scheme which is used in the following. It will be pointed out that condition (56) remains essential even in this simpler case.

4.1 Motion Rate Control: Simplified Version of the General Control Scheme

The basic idea of the control simplification consists in trying to obtain that the task error approximately behaves like a first-order decoupled system, i.e:

$$\dot{e} = -\lambda\, e \tag{67}$$

We also have:

$$\dot{e} = \frac{\partial e}{\partial \bar{r}}\, T + \frac{\partial e}{\partial t} \tag{68}$$

where the velocity screw T relies to $\dot{q}$ through:

$$\dot{q} = \left(\frac{\partial \bar{r}}{\partial q}\right)^{-1} T \tag{69}$$

$\left(\frac{\partial \bar{r}}{\partial q}\right)$ being the robot Jacobian matrix. We assume that the setting variable tunable by the user is simply the desired joint velocity, $\dot{q}_c$, as for most of industrial robots. Then, if it is possible to ensure that $\dot{q}_c \simeq \dot{q}$ (for example by using a large gain control, or when accelerations and disturbances are not too large), the velocity setting variable in se_3, that is to say the *desired* velocity screw T_c, may be considered as a pseudo–control vector. Its general form is:

$$T_c = \left(\widehat{\frac{\partial e}{\partial \bar{r}}}\right)^{-1} \left(-\lambda\, e - \widehat{\frac{\partial e}{\partial t}}\right) \tag{70}$$

Using (70) in (68) then gives:

$$\dot{e} = -\lambda \frac{\partial e}{\partial \bar{r}}\left(\widehat{\frac{\partial e}{\partial \bar{r}}}\right)^{-1} e - \frac{\partial e}{\partial \bar{r}}\left(\widehat{\frac{\partial e}{\partial \bar{r}}}\right)^{-1}\widehat{\frac{\partial e}{\partial t}} + \frac{\partial e}{\partial t} \tag{71}$$

which would give the evolution (67) if the models were perfect. Assuming in (71) that $\frac{\partial e}{\partial t} = \widehat{\frac{\partial e}{\partial t}} = 0$ (which is not strictly necessary for the analysis [24]), we can show that condition (56) suffices to ensure that $\|e\|$ decreases.

Let us now come back to the task function given by (65). Since s_d, W and C are chosen constant, we have:

$$\frac{\partial e}{\partial t} = W^+ C \frac{\partial s}{\partial t} + \beta\,(\mathbb{I}_6 - W^+ W)\frac{\partial}{\partial t}\left(\frac{\partial h_s}{\partial \bar{r}}\right)^T \tag{72}$$

Vector $\frac{\partial s}{\partial t}$ represents the contribution of a possible autonomous target motion and is often unknown. The choice made in the following is $\widehat{\frac{\partial s}{\partial t}} = 0$. If the target moves, this choice leads to a tracking error, the size of which decreases with λ (In [7], a simple estimation scheme of $\frac{\partial s}{\partial t}$ is proposed such that this tracking error progressively vanish for a constant velocity of the target). However, if, as in trajectory tracking, $\frac{\partial}{\partial t}\left(\frac{\partial h_s}{\partial \bar{r}}\right)$, is known, we may choose:

$$\frac{\widehat{\partial e}}{\partial t} = \beta \left(\mathbf{I}_6 - W^+ W\right)\frac{\partial}{\partial t}\left(\frac{\partial h_s}{\partial \bar{r}}\right)^T \tag{73}$$

Finally, it is assumed that the selected interaction models allow to satisfy condition (59) around the goal configuration, and that β is 'small enough' in the used hybrid tasks of form (65). The choice $\frac{\widehat{\partial e}}{\partial \bar{r}} = \mathbf{I}_6$ may therefore be done. Consequently, the simple form of the set velocity screw (i.e. used as the input of a low-level large gain control) is:

$$T_c = -\lambda\, e - \beta \left(\mathbf{I}_6 - W^+ W\right)\frac{\partial}{\partial t}\left(\frac{\partial h_s}{\partial \bar{r}}\right)^T \tag{74}$$

4.2 Simulation Results

In this section, simulation results related to the realization of two tasks, respectively performed from points (positioning) and lines (road following) are presented. Several other cases are predented in [6].

Positioning

Let us suppose that it is wished to set the camera with respect to a plane object which may be characterized by four points defining a square. Adequate sensor signals for this positioning task are the image coordinates of the four points: $s = (X_1, .., X_4, Y_1, ..Y_4)$. If the goal location is such that the image plane is parallel to the object plane with the four image points forming a centered square, the image feature is $s_d = (-a, a, a, -a, a, a, -a, -a)$ where $a = l/2z_d$, l being the vertex length and z_d the final wished range. The associated interaction matrix is easily obtained through (27):

$$L^T_{|s=s_d} = \begin{pmatrix} -1/z_d & 0 & -a/z_d & -a^2 & -1-a^2 & a \\ -1/z_d & 0 & a/z_d & a^2 & -1-a^2 & a \\ -1/z_d & 0 & a/z_d & -a^2 & -1-a^2 & -a \\ -1/z_d & 0 & -a/z_d & a^2 & -1-a^2 & -a \\ 0 & -1/z_d & a/z_d & 1+a^2 & a^2 & a \\ 0 & -1/z_d & a/z_d & 1+a^2 & -a^2 & -a \\ 0 & -1/z_d & -a/z_d & 1+a^2 & a^2 & -a \\ 0 & -1/z_d & -a/z_d & 1+a^2 & -a^2 & a \end{pmatrix} \tag{75}$$

It is also possible to compute $L^{T+}_{|s=s_d}$:

$$L^{T+}_{|s=s_d} = \begin{pmatrix} z_d\, c_1 & z_d\, c_1 & z_d\, c_1 & z_d\, c_1 & -z_d\, c_2 & z_d\, c_2 & -z_d\, c_2 & z_d\, c_2 \\ -z_d\, c_2 & z_d\, c_2 & -z_d\, c_2 & z_d\, c_2 & z_d\, c_1 & z_d\, c_1 & z_d\, c_1 & z_d\, c_1 \\ -z_d\, c_3 & z_d\, c_3 & z_d\, c_3 & -z_d\, c_3 & z_d\, c_3 & z_d\, c_3 & -z_d\, c_3 & -z_d\, c_3 \\ -c_4 & c_4 & -c_4 & c_4 & 0 & 0 & 0 & 0 \\ 0 & 0 & 0 & 0 & c_4 & -c_4 & c_4 & -c_4 \\ c_3 & c_3 & -c_3 & -c_3 & c_3 & -c_3 & -c_3 & c_3 \end{pmatrix} \tag{76}$$

$$\text{with } \begin{cases} c_1 = -1/4 \ , \ c_2 = (1+a^2)/4a^2 \\ c_3 = 1/8a \ , \ c_4 = 1/4a^2 \end{cases}$$

The matrix $L_{|s=s_d}^T$ is of full rank, which authorizes to choose the identity matrix $\mathbb{I}_6$ as W. The combination matrix is then $C = L_{|s=s_d}^{T+}$, which leads, owing to equation (65), to the following expression for the task function e to be regulated according to $T_c = -\lambda e$:

$$e = e_1 = L^{T+}(s(\bar{r},t) - s_d) \tag{77}$$

Figure 3 gathers some simulation results related to this example. Left and right top windows show respectively initial and final locations of the camera (symbolized by a pyramid) with respect to the target. Left and right middle windows represent the associated images. On left and right bottom windows, the time variation of $\|s(\bar{r},t) - s_d\|$ and of the components of T_c are respectively plotted. In this example, the value of λ is set to 0.1.

The exponential decreasing of $\|s - s_d\|$ and the convergence of the control law are always ensured even for an initial location of the camera far away from the desired one. Furthermore, the vertical line plotted on the curves at iteration 50 shows the time from which the positivity condition (56) is satisfied. These results emphasize the sufficiency of the condition (the convergence is ensured although the condition is not initially respected) and the good behavior of the control law with choosing $\hat{L} = L_{|s=s_d}$ and $C = W\hat{L}^{T+}$.

'Road' Following

Let us now consider a task whose aim is to position a camera with respect to a 'road', which is symbolized by three parallel straight lines in a plane (lateral and central white bands). The goal configuration is such that:

- the camera lies at a height y_d at the middle of the right lane;
- the camera axis z coincides with its direction and its axis y is vertical.

By using equations (28) and (30), the desired functions $h(\bar{x},\bar{p}_d)$ and $g(\bar{X},\bar{P}_d)$ associated to the three lines are immediately obtained:

$$h_1(\bar{x},\bar{p}_d) : \begin{cases} y + y_d = 0 \\ x + l/4 = 0 \end{cases} \Rightarrow \begin{cases} \theta_{1_d} = \arctan(-l/4y_d) \\ \rho_{1_d} = 0 \end{cases} \tag{78}$$

$$h_2(\bar{x},\bar{p}_d) : \begin{cases} y + y_d = 0 \\ x - l/4 = 0 \end{cases} \Rightarrow \begin{cases} \theta_{2_d} = \arctan(l/4y_d) \\ \rho_{2_d} = 0 \end{cases} \tag{79}$$

$$h_3(\bar{x},\bar{p}_d) : \begin{cases} y + y_d = 0 \\ x - 3l/4 = 0 \end{cases} \Rightarrow \begin{cases} \theta_{3_d} = \arctan(3l/4y_d) \\ \rho_{3_d} = 0 \end{cases} \tag{80}$$

where l is the width of the 'road'. The sensor signals to be selected for describing this task are the parameters which represent the three lines: $s =$

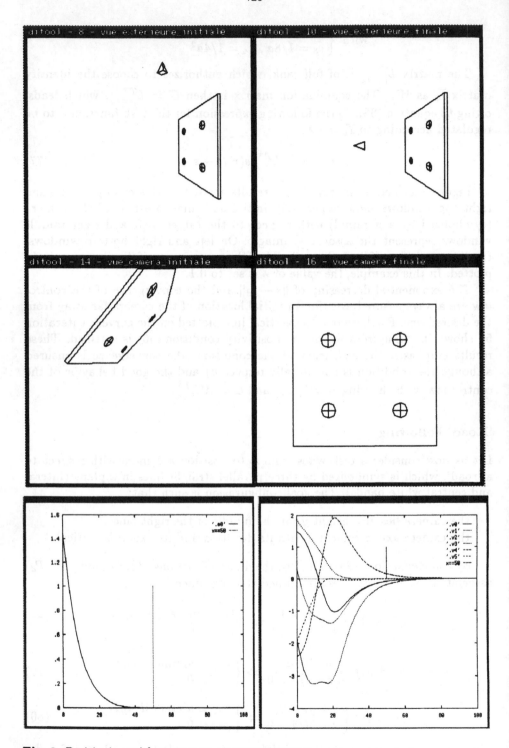

Fig. 3. Positioning with respect to a square target

$(\theta_1, \rho_1, \theta_2, \rho_2, \theta_3, \rho_3)$. Therefore $s_d = (\theta_{1_d}, 0, \theta_{2_d}, 0, \theta_{3_d}, 0)$. Furthermore, the interaction matrix associated with s_d may be easily derived from (35):

$$
L^T_{|s=s_d} = \begin{pmatrix}
-\cos^2\theta_{1_d}/y_d & -\cos\theta_{1_d}\sin\theta_{1_d}/y_d & 0 & 0 & 0 & -1 \\
0 & 0 & 0 & \sin\theta_{1_d} & -\cos\theta_{1_d} & 0 \\
-\cos^2\theta_{2_d}/y_d & -\cos\theta_{2_d}\sin\theta_{2_d}/y_d & 0 & 0 & 0 & -1 \\
0 & 0 & 0 & \sin\theta_{2_d} & -\cos\theta_{2_d} & 0 \\
-\cos^2\theta_{3_d}/y_d & -\cos\theta_{3_d}\sin\theta_{3_d}/y_d & 0 & 0 & 0 & -1 \\
0 & 0 & 0 & \sin\theta_{3_d} & -\cos\theta_{3_d} & 0
\end{pmatrix}
\tag{81}
$$

$L^T_{|s=s_d}$ is of rank 5, and $S^* = \mathrm{Ker}\, L^T_{|s=s_d} = (0\ 0\ 1\ 0\ 0\ 0)^T$. As expected, the considered image feature is invariant with respect to a translational motion along z and constitutes a virtual linkage of class 1.

Remarks:

- If a mobile robot, with translations along x, z and rotation around y as degrees of freedom, is used, the same task may be realized with a single straight line: $s = (\theta_1, \rho_1)$.
- Ambiguity of the (ρ, θ) representation, evoked in section 2.2.4, which concerns the possible choices for θ ($\theta + 2k\pi, \forall k \in \mathbb{N}$), is overcome by modulating to 2π the error $(\theta - \theta_d)$.

Let us now apply the approach of section 3.1.3 for the derivation of e. The following 5×6 matrix may be chosen as a matrix W:

$$
W = \begin{pmatrix}
1 & 0 & 0 & 0 & 0 & 0 \\
0 & 1 & 0 & 0 & 0 & 0 \\
0 & 0 & 0 & 1 & 0 & 0 \\
0 & 0 & 0 & 0 & 1 & 0 \\
0 & 0 & 0 & 0 & 0 & 1
\end{pmatrix}
\tag{82}
$$

The combination matrix C is chosen equal to $WL^{T+}_{|s=s_d}$ and, by using (65), the following task vector e is obtained:

$$
e = W^+ W L^{T+}_{|s=s_d}(s(\bar{r}, t) - s_d) + \beta\left(\mathbb{I}_6 - W^+ W\right)\frac{\partial h_s}{\partial \bar{r}}^T
\tag{83}
$$

The secondary task may consist in specifying a time trajectory along z, for example a constant velocity V. The associated secondary cost to be minimized is $h_s = \frac{1}{2}(z(t) - z_0 - Vt)^2$ with $z(0) = z_0$. Therefore:

$$
\frac{\partial h_s}{\partial \bar{r}} = (0\ \ 0\ \ (z(t) - z_0 - Vt)\ \ 0\ \ 0\ \ 0)
\tag{84}
$$

Note that tasks e_1 and $e_2 = z(t) - z_0 - Vt$ are then compatible and independent since:

$$e = \begin{pmatrix} 1\,0\,0\,0\,0\,0 \\ 0\,1\,0\,0\,0\,0 \\ 0\,0\,0\,0\,0\,0 \\ 0\,0\,0\,1\,0\,0 \\ 0\,0\,0\,0\,1\,0 \\ 0\,0\,0\,0\,0\,1 \end{pmatrix} L^{T+}_{|s=s_d} \left(s(\bar{r},t) - s_d \right) + \beta \begin{pmatrix} 0 \\ 0 \\ z(t) - z_0 - Vt \\ 0 \\ 0 \\ 0 \end{pmatrix} \tag{85}$$

T_c is obtained from equation (74), in which we have now $\frac{\partial}{\partial t}\left(\frac{\partial h_s}{\partial \bar{r}}\right) = (0\ 0\ -V\ 0\ 0\ 0)$.

Simulation results of figure 4 are structured as in the previous case. Parameters are tuned with the following values: $\lambda = 0.1, \beta = 1$ and $V = 1.25\text{cm/s}$.

4.3 Experiment With a Six-jointed Robot

The positioning task with respect to a square target was implemented on an experimental testbed including a CCD camera mounted in the effector of a six-jointed robot. A camera calibration step allowing to obtain the model of figure 1 was done, and the transformation matrix from the camera frame to a frame linked to the end effector was identified using methods given in [5]. The computation of the inverse robot Jacobian matrix is realized on a 68020-based dedicated board with a sampling rate of 5 ms. The useful part of the scene is a set of four white disks on a dark background. Owing to this simplicity, mass centers of image disks and velocity screw T_c are computed in less than 20 ms, which ensures that the video rate is respected. Thus, visual features of all odd frames are taken into account in the control. For more complex scenes, requiring a vision process time higher than 20 ms, an asynchronous feature data extraction could be used as illustrated in [14].

Figure 5 shows a sequence of six images taken during a positioning task. The used value of λ is 0.1. The plottings of $\|s - s_d\|$ and T_c are close to the one obtained in simulation.

Let us finally point out that, when the object motion is slow enough to ensure that it keeps lying inside the camera field from an iteration to the following, a tracking error decreasing with λ is observed (cf section 4.1). The stability of the system is however not affected. This error may be reduced by increasing the gain λ (which has not to be set too high in order to preserve the stability of the system) or by introducing in T_c an estimation of the object velocity.

5 Conclusion

The objective of this paper was to show that technical progress now permits to control interaction between a robot and its environment by using data provided by a visual sensor directly inside a closed-loop control scheme. When exploiting such an approach in robotics, two questions need to be answered:

132

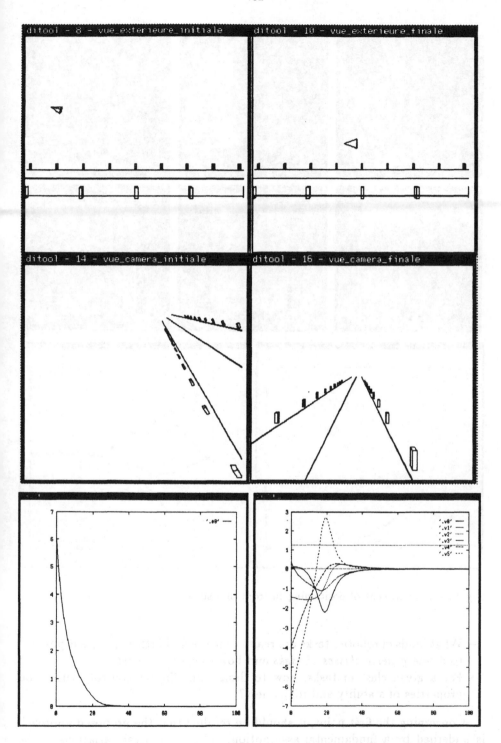

Fig. 4. Road following

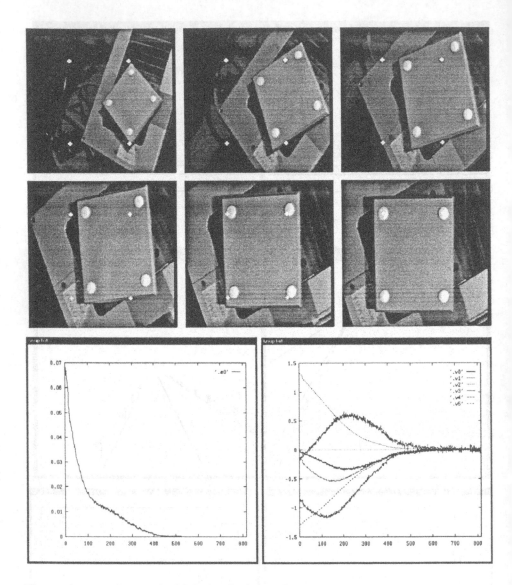

Fig. 5. An experiment of positioning in front of a square

- What kinds of robotics tasks are really concerned? Furthermore, is it possible to define *generic classes of tasks* and how to specify them?
- For a given class of tasks, how to design an efficient control, with good properties of stability and robustness?

Concerning the first point, it should be recalled that the proposed approach is underlied by a fundamental assumption: only geometrical variations of the environment are liable to make the sensor output vary. From that, it appears that the tasks to be considered are those which require the control of geometrical interaction between a robot and its environment. Typically, it will be wished to control the location and the attitude of a frame linked to a sensor with regard to a frame linked to the environment. A mechanics-based formalism may then be used, like in assembly applications. Tasks may be specified as basic *linkages* (point on plane, line on plane...) which represent in a simple way the aimed constraints and the available degrees of freedom. This formalism, extended to non-contact sensor-based tasks allows to define elementary task classes (positioning, line following...) to be combined in more complex applications. In the *visual servoing* approach, we have established for several simple scene primitives (point, line, circle, sphere...) structure and properties of interaction screws, allowing in that way to associate a linkage under the above formalism with a scene feature. Because of the nature itself of a visual sensor, the main encountered difficulty comes from the infinite number of possibilities which exist for relating a given type of linkage to an image feature. Preferring a feature to another depends therefore on other criteria, related both to image processing aspects (robustness or implementation easiness of feature extraction algorithms) and to control requirements (sensitivity, decoupling...). Much work remains to be done in that area.

The second point, which concerns the implementation of robust control schemes, has already been deeply investigated in the general framework of robot manipulator control. In the case of visual servoing, it remains to take into account the possibility of tracking moving objects with unknown trajectories. Since the use of large gains is limited in practice because of sampling rates, the inclusion in the control of an algorithm estimating on line the relevant parameters of the motion of a target seems to be necessary.

In conclusion, let us emphasize that the implementation of visual-based servo control loops requires some coherence between the sampling frequency at the image level and the bandwith of the system to be controlled. Although the features used in the proposed approach are low-level and that no semantical interpretation is needed, it seems now still necessary to use dedicated real-time architectures for the primitive extraction step. Expected technical progress will however surely enable the application fields of this approach to be extended in the next years.

References

[1] G. J. Agin, "Real Time Control of a Robot with a Mobile Camera", *SRI International*, Technical Note 179, February 1979.

[2] G. André, R. Fournier, "Generalized End Effector Control in a Computer Aided Teleoperation System with Application to Motion Coordination of a Manipulator Arm on a Oscillating Carrier", *Int. Conf. on Advanced Robotics*, Tokyo, Japan, pp. 337-344, September 1985.

[3] D.J. Balek, R.B. Kelley, "Using Gripper Mounted Infrared Proximity Sensors for Robot Feedback Control", *IEEE Int. Conf. on Robotics and Automation*, Saint Louis, Missouri, USA, pp. 282-287, March 1985.

[4] P. Bouthemy, "A Maximum Likelihood Framework for Determining Moving Edges", *IEEE Trans. on Pattern Analysis and Machine Intelligence*, Vol. 11, n. 5, pp. 499-511, May 1989.

[5] F. Chaumette, P. Rives, "Réalisation et calibration d'un système expérimental de vision composé d'une caméra embarquée sur un robot manipulateur", *INRIA Research Report*, n. 994, March 1989.

[6] F. Chaumette, "La relation vision-commande : théorie et applications à des tâches robotiques", *Ph-D Thesis*, Rennes I University, France, July 1990.

[7] F. Chaumette, P. Rives, B. Espiau, "Positioning of a Robot with respect to an Object, Tracking it and Estimating its Velocity by Visual Servoing", *IEEE Int. Conf. on Robotics and Automation*, Sacramento, USA, Vol. 3, pp. 2248-2253, April 1991.

[8] E. Cheung, V. Lumelsky, "Proximity Sensing in Robot Manipulator Motion Planning: System and Implementation Issues", *IEEE Trans. on Robotics and Automation*, Vol. 5, n. 6, pp. 740-751, December 1989.

[9] P. I. Corke, R. P. Paul, "Video-Rate Visual Servoing for Robots", *First Int. Symposium on Experimental Robotics*, Montreal, Canada, June 1989.

[10] B. Espiau, R. Boulic, "Collision Avoidance for Redundant Robots with Proximity Sensors", *Third Int. Symposium on Robotics Research*, Gouvieux, pp. 243-251, MIT Press, Cambridge, October 1985.

[11] B. Espiau, P. Rives, "Closed-Loop Recursive Estimation of 3D Features for a Mobile Vision System", *IEEE Int. Conf. on Robotics and Automation*, Raleigh, North Carolina, USA, Vol. 3, pp. 1436-1443, April 1987.

[12] B. Espiau, "Sensory-based Control: Robustness Issues and Modeling Techniques; Application to Proximity Sensing", *NATO Advanced Research Workshop on Kinematic and Dynamic Issues in Sensor-based Control*, Il Ciocco, Italy, pp. 3-44, October 1987.

[13] J. T. Feddema, C. S. G. Lee and O. R. Mitchell, "Automatic selection of image features for visual servoing of a robot manipulator", *IEEE Int. Conf. on Robotics and Automation*, Scottsdale, Arizona, USA, Vol. 2, pp. 832-837, May 1989.

[14] J. T. Feddema, O. R. Mitchell, "Vision-Guided Servoing with Feature-Based Trajectory Generation", *IEEE Trans. on Robotics and Automation*, Vol. 5, n. 5, pp. 691-700, October 1989.

[15] C. L. Fennema, W. B. Thomson, "Velocity Determination in Scenes Containing Several Moving Objects", *Computer Graphics and Image Processing*, Vol. 9, pp. 301-315, 1979.

[16] A. Flynn, "Combining Sonar and Infrared Sensors for Mobile Robot Navigation", *Int. Journal of Robotics Research*, Vol 7, n. 6, pp. 5-14, December 1988.

[17] A. L. Gilbert and al., "A Real-Time Video Tracking System", *IEEE Trans. on Pattern Analysis and Machine Intelligence*, Vol. 2, n. 1, pp. 47-56, January 1980.

[18] M. Kabuka, E. McVey, P. Shironoshita, "An Adaptive Approach to Video Tracking", *IEEE Journal of Robotics and Automation*, Vol. 4, n. 2, pp. 228-236, April 1988.

[19] P. Rives, "Dynamic vision: theoretical capabilities and practical problems", *NATO Workshop on Kinematic and Dynamic Issues in Sensor Based Control*, Italy, pp. 251-280, October 1987.

[20] P. Rives, F. Chaumette, B. Espiau, "Visual Servoing Based on a Task Function Approach", *First Int. Symposium on Experimental Robotics*, Montréal, Canada, June 1989.

[21] C. Samson, "Une approche pour la synthèse et l'analyse de la commande des robots manipulateurs", *INRIA Research Report*, n. 669, May 1987.

[22] C. Samson, B. Espiau, M. Le Borgne, "Robot Redundancy: an Automatic Control Approach", *NATO Advanced Research Workshop on Robots with Redundancy*, Salo, Italia, June 1988.

[23] C. Samson, B. Espiau, "Application of the Task Function Approach to Sensor-Based-Control of Robot Manipulators", IFAC, Tallin, USSR, July 1990.

[24] C. Samson, B. Espiau, M. Le Borgne, *Robot Control: the Task Function Approach*. Oxford University Press, 1991.

[25] A. C. Sanderson, L. E. Weiss, "Image Based Visual Servo Control Using Relational Graph Error Signal", *Int. Conf. on Cybernetics and Society*, Cambridge, MA, IEEE SMC, pp 1074-1077, October 1980.

[26] A. C. Sanderson, L. E. Weiss, "Adaptive Visual Servo Control of Robots", Reprinted in *Robot Vision*. A. Pugh, Ed. Bedford, UK:IFS Pub. Ltd., pp. 107-116, 1983.

[27] M. Spivak, *A comprehensive introduction to differential geometry*. Publish or Perish, Boston, 1970.

[28] C. Wampler, "Multiprocessor Control of a Telemanipulator with Optical Proximity Sensors", *Int. Journal of Robotics Research*, Vol. 3, n. 1, pp. 40-50, 1984.

[29] L. E. Weiss, "Dynamic Visual Servo Control of Robots. An Adaptive Image based Approach", *Technical Report*, CMU-RI-TR-84-16; Carnegie Mellon, 1984.

[30] L. E. Weiss, A. C. Sanderson, "Dynamic Sensor-Based Control of Robots with Visual Feedback", *IEEE Journal of Robotics and Automation*, Vol. 3, n. 5, pp. 404-417, October 1987.

Geometric Tools
for
Visual Perception

Geometric Solutions
to some 3D Vision Problems

Roger Mohr and Luce Morin

LIFIA–IMAG
46 AV. FÉLIX VIALLET,
38031 GRENOBLE, FRANCE

Abstract: In this paper, we give geometric constructive solutions to the 3D vision problems such as

- how to back project an image point and how to project a space point in monocular vision.
- how to determine the euclidean geometry in stereo vision.
- how to locate the optical center of the cameras.

No direct calibration is needed, the necessary reference points are directly included in the computation. The method involves only simple geometric computation. The appropriate mathematical tool, projective geometry, is introduced. Results on preliminary experiments are discussed and some future research directions are indicated.

1 Introduction: 3D Vision Problems

In a chapter surveying projective geometry, Duda and Hart [6] explicitly mention the current lack of understanding on the subject. This paper tries to provide a better understanding of this problem and studies projective geometry with the goal of deriving 3D relative positioning of objects in a scene. This is a key point of scene understanding since it allows to reason on positions without knowing the exact numerical place of the coordinates or size of the observed items.

1.1 The 3D Perception

In 3D passive vision, cameras are represented with a pinhole model. It is central projection. So each of the point observed on a image corresponds to an optical line which passes through the optical center. Therefore in order to reconstruct the 3D structure of the observed scene, a vision system must first be able to determine the viewing line of the point observed on the image plane. This is what is called the inverse perspective problem.

* This work is supported by Bull, Esprit project Ivica, Orb ... 3D ... and the Pôle Productique Rhône-Alpes.
L. Morin has presently a fellowship from Thomson ...
It is besides a pleasure to acknowledge R. Horaud, T. Collin ... Wachter and ... E. Arbogast for their fruitful remarks.

Geometric Solutions
to some 3D Vision Problems

Roger Mohr and Luce Morin

LIFIA–IRIMAG
46 AV. FÉLIX VIALLET,
38031 GRENOBLE, FRANCE

Abstract. In this paper, we give geometric constructive solutions to the 3D vision problems such as

- how to back project an image point and how to project a space point in monocular vision,
- how to determine the epipolar geometry in stereo vision,
- how to locate the optical center of the camera.

No direct calibration is needed, the necessary reference points are directly included in the computation. The method involves only simple geometric computation. The appropriate mathematical tool, projective geometry, is introduced. Results on preliminary experiments are discussed and some future research directions are indicated.

1 Introduction: 3D Vision Problems

In a chapter surveying projective geometry, Duda and Hart [6] explicitly mention the current lack of understanding on the subject. This paper tries to provide a better understanding of this problem and studies projective geometry with the goal of deriving 3D relative positioning of objects in a scene. This is a key point of scene understanding and it allows to reason on positions without knowing the exact numerical values of the coordinates or size of the observed items.

1.1 The 3D Perception

In 3D passive vision, cameras are represented with a pinhole model, i.e. central projection. So each of the point observed on a image correspond to an ideal line which passes through the optical center. Therefore, in order to reconstruct the 3D structure of the observed scene a vision system must first be able to *determine the viewing line* of the point observed on the image plane. This is also what is called the inverse perspective problem.

[0] This work is supported by BRA FIRST project from CEC DG XIII and the GRECO PRC Communication Homme–Machine under the project ORASIS.

L. Morin has presently a fellowship from Greco Traitement du Signal et des Images.

It is also a pleasure to acknowledge R. Horaud, Y. Colin de Verdières and particularly E. Arbogast for their fruitful remarks.

Dealing with two cameras in stereo vision the fundamental geometric problem is the determination of the locus in image 2 of the possible matches of one point in image 1: this is the problem of *epipolar geometry* and will be addressed in Sect.3.5.

Other problems addressed within this paper includes the *location of the camera center* and how to derive some general geometric properties of the epipolar geometry.

1.2 Calibration

In order to complete 3D visual tasks it is usually necessary to perform calibration of the vision system. Once the camera parameters are known, the problem of the 3D reconstruction becomes much easier (see for instance [9], [17], [16] for recent contributions).

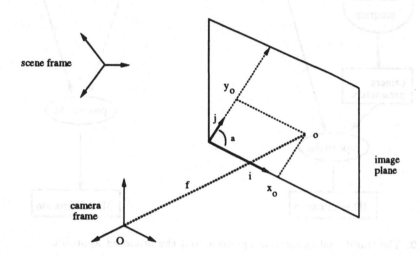

Fig. 1. The intrinsic parameters

Calibration requires the computation of five intrinsic parameters (see Fig.1):

- the two coordinates x_o and y_o of o, the projection of the optical center O in the image plane,
- the two units, i and j on the frame axes of the image plane and the focal distance f (up to a common scale factor, which makes only two independent parameters),
- and the angle a between these two axes (if they are not perpendicular).

To associate the camera related frame to a fixed world frame, six extrinsic parameters are needed:

- three for a pure rotation,
- and three for a pure translation.

The usual approach to these problems in computer vision is either to calibrate explicitly each of intrinsic and extrinsic parameters of the camera or to model the whole imaging system by a 4 × 3 projection matrix (12 parameters to be determined up to a common scalar factor, that is, only 11 independent parameters) in homogeneous coordinates (see Sect.2). The first process involves often a tedious and instable computation (see [8, 2], [3]) while the second is much easier and often sufficient.

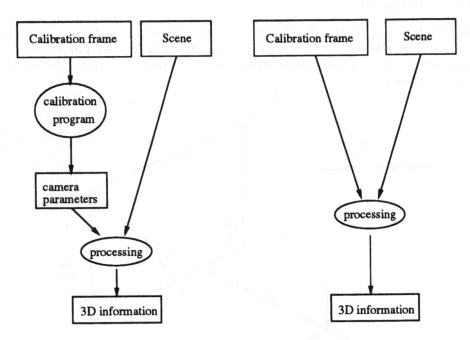

Fig. 2. The traditional calibration approach and the proposed approach

However the main problem in vision is to refer what we see to some reference points, as for instance when we want to know if the robot is far from a table. We show here that solutions to these problems can be obtained by direct and simple geometric constructions using projective properties. No direct calibration is needed, only scene reference points are necessary to back project image points and to solve some other problems like finding the epipolar lines in stereovision. Figure 2 depicts our approach in contrast with the standard one : instead using references points for computing the camera parameters which are afterwards used for measures, the reference points are directly introduced in the measures.

1.3 Outline of the Paper

Projective geometry [12] is our central mathematical tools, since projective geometry deals with the properties that are invariant under homography and therefore under central projection for uncalibrated cameras. When the camera parameters are settled —which is not the case in our framework— the problem addressed here becomes much easier and the retrieved positions are referred to an absolute reference frame, whereas our method leads to relative positioning.

Sparr and Nielsen give an alternative approach to the one provided here [13]. Using a slight generalization of barycentric coordinates, they deduce an invariant under central projection on these coordinates. It allows them to deduce some of results we present here as well. In both approaches the tools belong to the same family, but they have to be handled in very different ways.

Another essential aspect of this geometric reconstruction approach is that it allows to better understand the calibration behavior. It gives us an intuitive display of the imaging, projection and back projection process. Therefore it is possible to find more stable and robust methods.

Next section introduces briefly the geometric tools needed to establish our main results: cross-ratio and Chasles' theorem. Besides, it defines "projective coordinates", a way to reference a point in a plane with projective invariant values.

Section 3 shows how to use these tools to solve the main problems we address here but considering only the case where at least four of the reference points are coplanar. Section 4 extends the solutions to the general case.

Some experimental results are presented in Sect.5. They illustrated the robustness of the method. The last section discuss some extension of the present work. In particular it will be shown how our tools allow to make relative positioning using reference objects whose dimensions are unknown.

2 Few Results in Projective Geometry

2.1 Preliminary Definitions: Projective Space and Homogeneous Coordinates

We provide here a short introduction to projective geometry definitions and vocabulary. The reader is referred to [12] for a gentle introduction or to [10] for vision oriented considerations on projective geometry. This subsection is helpful but not essential for the comprehension of the remainder.

We consider $I\!R^{n+1} - \{(0, ..., 0)\}$ with the equivalence relation:

$$(x_1, \ldots, x_{n+1}) \sim (x'_1, \ldots, x'_{n+1}) \ \exists \lambda \neq 0 \text{ such that } (x'_1, \ldots, x'_{n+1}) = \lambda(x_1, \ldots, x_{n+1}).$$
$$(1)$$

The quotient space obtained by this equivalence relation is the projective space $I\!P^n$. Thus the $n+1$-tuples of coordinates $(x_1, ..., x_{n+1})$ and $(x'_1, ..., x'_{n+1})$ represent the same point in the projective space.

A projective basis is a set of $n+1$ points of IP^n which are linearly independent (for example, the set $e_i = (0, \ldots, 1, \ldots, 0))$. Any point x can be described as a linear combination of the basis:

$$x = \sum_{i=1}^{i=n+1} a_i e_i \qquad (2)$$

The $(n + 1)$-tuples $(\lambda a_i)_{i=1\ldots n+1}$, $\lambda \neq 0$, satisfying this equality, are called homogeneous coordinates of x in the basis $(e_i)_{i=1\ldots n+1}$.

Any linear transformation in homogeneous coordinates is a homography. The matrix associated with a given homography is defined up to a non zero scale factor, that is $\lambda y = Ax$. Notice that a projection from the 3D space IP^3 onto the image plane IP^2 is therefore represented by a 4×3 matrix in homogeneous coordinates which leads us to a 11-dimensional space for these projections: exactly the same dimension as the space of all possible projections for uncalibrated camera, as stated in the introduction.

The usual affine space IR^n is mapped into IP^n by the correspondence Ψ:

$$\Psi : (x_1, \ldots, x_n) \rightarrow (x_1, \ldots, x_n, 1). \qquad (3)$$

Notice that only the points $(y_1, \ldots, y_n, 0)$ are not reached by Ψ. These points are considered as points at infinity. They may be perceived as the limit of

$$(y_1, \ldots, y_n, \lambda) \sim (y_1/\lambda, \ldots, y_n/\lambda, 1) \qquad (4)$$

while $\lambda \rightarrow 0$.

2.2 The Basic Invariant

The cross-ratio is the basic invariant in projective geometry: all other projective invariants can be derived from it [7]. It has to be pointed out that the negative results found by Burns and al [4] is not applicable : they just prove that there is no invariant for projective correspondence; but projective correspondence is not a homographic mapping (it is not a mapping at all). So projective correspondence is not the right tool to be used.

- Cross-ratio of four points:
  Let A, B, C, D be four collinear points, we can define what we call their *cross-ratio* as:

$$[A, B, C, D] = \frac{\overline{CA}}{\overline{CB}} / \frac{\overline{DA}}{\overline{DB}} \qquad (5)$$

where $\overline{AB}$ is the algebraic measure of AB.

The fundamental theorem about cross-ratios follows :

Theorem. Any homography preserves the cross-ratio.

A homographic transformation is any linear transformation in homogeneous coordinates, thus including central projection (perspective projections), linear scalings, skewings, rotations, translations. . . .

The invariance of the cross-ratio is illustrated by Fig.3, and is expressed by:

$$[A, B, C, D] = [A', B', C', D'].\qquad(6)$$

– Cross-ratio of four lines

The cross-ratio of a pencil of four lines l_1, l_2, l_3, l_4 going through O is defined as the cross-ratio $[A, B, C, D]$ of the points of intersection of the l_i with any line l not passing through O.

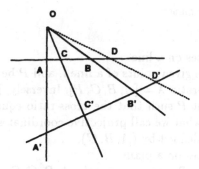

Fig. 3. Cross-ratio of four lines

This is, of course, independent of the choice of l. As we know, this cross-ratio can also be defined as:

$$\frac{\sin(COA)}{\sin(COB)} \Big/ \frac{\sin(DOA)}{\sin(DOB)}\qquad(7)$$

which has nothing to do with the choice of the line l; this point is illustrated by Fig.3.

– Cross-ratio of four planes

The cross-ratio of a pencil of four planes p_1, p_2, p_3, p_4 is defined as the cross-ratio $[l_1, l_2, l_3, l_4]$ of their four lines of intersection with any plane p. This is, of course, independent of the choice of p (see figure 4).

2.3 Projective Coordinates

We define here how points can be designated through "coordinates" with respect to given reference points using cross-ratio. Obviously from the previous section these "coordinates" are invariant under any homography, and therefore under central projection and scaling along the image plane axes.

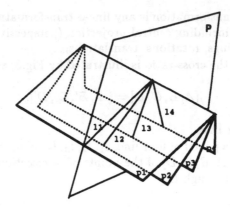

Fig. 4. Cross-ratio of four planes

- Projective coordinates on a line

  Let A, B, C be three given points of a line l, and P be a point of l. P and A, B, C define a cross-ratio $\lambda = [A, B, C, P]$. Inversely, given a scalar λ there exists a unique point P such that the cross-ratio equals to λ. (x_1, x_2) such that $x_1/x_2 = \lambda$ are what we call projective coordinates of the point P in the coordinates system defined by (A, B, C).

- Projective coordinates on a plane

  In a projective plane $\mathcal{P}$, any four points A, B, C, D, no three collinear, define a projective coordinates system (see Fig.5). Given a point P of $\mathcal{P}$, let (x_1, x_2, x_3) be a triple of real numbers defined up to a scaling factor and such that :

$$\frac{x_1}{x_2} = [CA, CB, CD, CP] \tag{8}$$

$$\frac{x_2}{x_3} = [AB, AC, AD, AP] \tag{9}$$

(x_1, x_2, x_3) are called the projective coordinates of P in the coordinates system (A, B, C, D). Naturally we also have $x_3/x_1 = [BC, BA, BD, BP]$.

In a projective plane with four known points we can uniquely reference any point of the plane by their projective coordinates so defined.

In fact only the two cross-ratios $k_1 = x_1/x_2$ and $k_2 = x_2/x_3$ are necessary to uniquely define a point; therefore they will be used throughout this paper instead of the projective coordinates.

2.4 Implementation

This section describes how to compute the projective coordinates from the geometric data in the image plane (or in a 3D plane) and how determine the geometric position of a point in a plane, given its projective coordinates. In other words it provides the computation of the correspondence between Cartesian and projective coordinates.

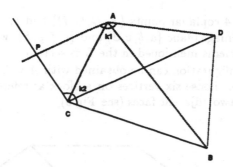

Fig. 5. Projective coordinates in the plane

Let A and B be two points in the 3D space, and let a^t and b^t be their Cartesian coordinates. The line going through A and B is the set of points M_μ whose coordinates are $(1 - \mu)a^t + \mu b^t$. So each point on the line is associated in one to one way with a real number μ and for $\mu = 0$ the point is equal to A. This representation is usually called the barycentric representation of the line.

The cross-ratio of four points $M_{\mu_i}, i = 1, .., 4$ is then:

$$[M_{\mu_1}, M_{\mu_2}, M_{\mu_3}, M_{\mu_4}] = \frac{\mu_3 - \mu_1}{\mu_4 - \mu_1} \times \frac{\mu_4 - \mu_2}{\mu_3 - \mu_2} \qquad (10)$$

In a plane, the intersection of two lines $L_{A,B}$ and $L_{C,D}$ is easily solved as a linear problem with two unknown λ and μ which satisfy:

$$(1 - \mu)a^t + \mu b^t = (1 - \lambda)c^t + \lambda d^t \qquad (11)$$

Using these tools, the cross-ratio of a pencil of lines can then be computed as the cross-ratio of the intersections of the pencil with any line. From there the cross-ratios of the two pencils which locate a point in the plane (see Fig.5) are directly obtained. These cross-ratios are the coordinates defined in Sect.2.3.

Working in the reverse way is also easy. First notice that knowing 3 points and the value of the cross-ratio, we obtain directly the μ associated to the forth unknown point from equation 10. Hence, from a pencil of lines with a given cross-ratio and 3 known lines, the remaining line is directly computed. Therefore the two cross-ratios defining a point in the plane allow to compute the dotted lines of Fig.5 and their intersection provides the Cartesian coordinates we are looking for.

3 Projective Reconstruction Using Coplanar Points

This section solves the problems mentioned in the introduction in the simplest case when at least four of the reference points are coplanar. First the back projection is solved and the technique introduced is used for solving the other ones.

Given two sets of 4 coplanar points $\{A, B, C, D\}$ and $\{E, F, G, H\}$ and their projections onto the image plane $\{a, b, c, d\}$ and $\{e, f, g, h\}$, we are going to show how to solve the problems mentioned in the first section.

Note that this configuration can be obtained with $A = E$ and $B = F$; this is the case each time we choose six vertices on a block as reference points, the six points lying on only two adjacent faces.(see Fig.6)

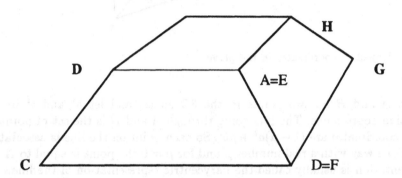

Fig. 6. 6 points on adjacent faces provide 2 sets of 4 coplanar points

3.1 How to Back Project an Image Point?

Given an observed point m on the image plane, it is possible to determine the location of the viewing line Om through m with respect to the scene, without any calibration.

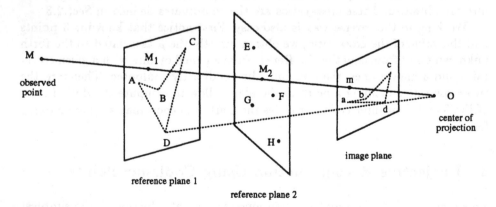

Fig. 7. The back projection of the image point m

Proof. As shown in Sect.2.4, we can determine the projective coordinates (x_1, x_2, x_3) of m with respect to the coordinate system (a, b, c, d), from measures in the image plane. Let M_1 be the intersection in 3D space of the viewing line Om with the plane $ABCD$ (see Fig.7). The projection of M_1 onto the image plane is m. Since the projective coordinates are invariant under projection (see Sect.2.3), the projective coordinates of M_1 with respect to (A, B, C, D) are (x_1, x_2, x_3). The position of M_1 in the reference plane $ABCD$ can then be determined from its projective coordinates as shown in Sect.2.4.

We can similarly determine M_2, the intersection of the viewing line Om with the plane $EFGH$. (M_1, M_2) determine the viewing line with respect to the scene, and this *without knowing the camera position in the scene.* ☐

3.2 Where is the Camera?

We just showed in 3.1 how to reconstruct the viewing line from a single shot. It is also possible to reconstruct the position of the optical center with respect to the scene, in a single shot, using the same method. If two points m and p are given in the image plane, it is possible to reconstruct the viewing line Om and Op. These two lines happen to intersect in space at O, which provides us with a method to compute the location of O.

Furthermore, if more than two points are available, it will be possible to derive a solution using the least squares estimate, which will be much more reliable against noise in the measures.

3.3 Where is the Image Point?

We suppose here that the position of camera center O is computed using the previous method. The image point m of a space point M is then easily constructed: First compute the intersection point of OM with a known plane containing four reference points; then in this known plane take measures of the projective coordinates of this intersection as defined in Sect.2.3. Finally with these two known cross-ratios construct the point m in the image plane using the image projection of the four reference points.

3.4 How to Locate a Point in the Scene?

The previous method can be applied to locate a point in a scene containing reference points. Here two snapshots of the same scene are needed. A given point M in the 3D space is projected respectively as m_1 and m_2 in each of these two images.

Let us suppose that the matching of m_1 and m_2 has been solved. For each image we are able to reconstruct the viewing line passing through M and therefore these two lines intersect on M. This method degenerates only when the two positions of the optical centers are aligned with the viewing line.

3.5 Where are the Epipolar Lines?

Finding one Epipolar Line In stereo vision, the epipolar centers o and o' are the intersection of the line OO' with the image planes (see Fig.8). These points are quite important as all epipolar lines are radiated from these centers. The epipolar constraint states that given a point M in 3D space with projections m in the first image plane and m' in the second one then the epipolar plane $OO'M$ intersects the image planes along the lines om and $o'm'$. Therefore given a point m in the first image plane, its image m' in the second image lies on the corresponding epipolar line. The line om (resp. $o'm'$) is what we call the epipolar line of the point m' (resp. m).

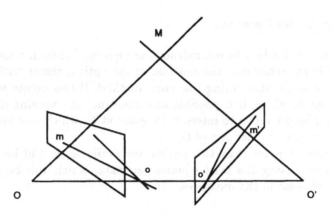

Fig. 8. The epipolar geometry. If an image plane does not intersect the base line OO', its epipolar center is at infinity in the direction of OO'.

The same method defined in 3.1 enables to determine the epipolar lines.

The two sets of 4 coplanar points $\{A, B, C, D\}$ and $\{E, F, G, H\}$ respectively project as $\{a, b, c, d\}$ and $\{e, f, g, h\}$ onto the image plane. Given an observed point m on the image plane, it is possible to determine the location of the viewing line Om through m with respect to the scene.

Another snapshot from a new optical center O' provides a new image. We suppose here that this image contains also references points (not necessary the same as for image 1, but within the same reference frame). It is then possible to determine the projection of the viewing line Om on the new image plane, which is the epipolar line of point m. Indeed, we know the projective coordinates of points $M1$ and $M2$ (as defined in 3.1), and we are therefore able to reconstruct points $m1$ and $m2$ on the new image plane. These two points determine the epipolar lines of point M on the new image plane.

The pencil of Epipolar Lines Projective geometry allows us to prove easily the following result which constrains the position of the epipolar lines.

Theorem. *Two pencils of four corresponding epipolar lines have same cross-ratio.*

Proof. From their definition, corresponding epipolar lines are the intersections of a 3D plane containing the base line OO' with the two image planes. Therefore the cross-ratio of the pencil of lines is the cross-ratio of the pencil of planes. □

This property could also be obtain as a consequence of Ayache's results [1] which stated that the epipolar lines correspond through a affine transformation

From there, if three corresponding epipolar lines are known, the epipolar geometry is completely defined: for any point m in image1, m defines the epipolar line going through m in image1. The cross-ratio of these known four lines can then be computed and we define the corresponding line in the second image as the one with the same cross-ratio.

4 Projective Reconstruction with non Coplanar Points

This section uses the technique developed by Trip [14] in the planar case and extends it in the 3D case. As this author we need Chasles theorem, which will be first introduced.

4.1 Chasles Theorem

Last century, Chasles proved the following theorem:

Theorem 1. *In the projective plane, let A, B, C, D be four points, not three of them collinear. The locus of the center of a pencil of line passing through these 4 points and having a given cross-ratio is a conic. Reciprocally, if M lies on a conic passing through A, B, C and D the cross-ratio of the four lines MA, MB, MC, MD is independent of the point M on the conic.*

Figure 9 illustrates this result: if P and Q are two points on a conic, we have

$$[PA, PB, PC, PD] = [QA, QB, QC, QD] \qquad (12)$$

This conic can easily be derived from the data. Let L_{AB} be the 1st degree polynomial equation of the line going through A and B.

$$L_{AB} = (x_A - x)(y_A - y_B) - (y_A - y)(x_A - x_B) \qquad (13)$$

Notice then that

$$C_\lambda = \lambda L_{AB} L_{CD} + (1 - \lambda) L_{AC} L_{BD} \qquad (14)$$

defines a second degree polynomial in x and y which has A, B, C and D as roots. C_λ defines exactly the family of conics passing through these four points. A direct computation proves that C_k is Chasles' conic for the cross-ratio k.

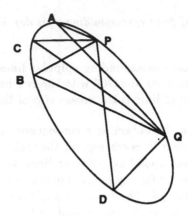

Fig. 9. Invariance of cross-ratio on a conic

4.2   Where is the Camera?

We consider a set of 6 points A, B, C, D, E, F in the 3D space and their projections a, b, c, d, e, f onto the image plane. Let us assume that the unknown optical center is O. Let us consider the pencil of planes going through OA and the points B, C, D, E, F respectively.

The intersection of these planes with the image plane are the lines ab, ac, ad, ae, af, thus we can measure the cross-ratio of any pencil made with 4 of these planes.

Any known plane p intersects this pencil along lines going through a', intersection of OA with p. The lines AB (resp: AC, AD, AE, AF) intersects the plane p in the known point b' (resp: $c', \ldots, f'$).

As we know the cross-ratio $[a'b', a'c', a'd', a'e']$, the point a' is on the conic determined by b', c', d', e' in the plane P according to Chasles's theorem. Similarly we can do the same construction with the point f' instead of e', thus a' lies on the conic determined by b', c', d', f' on P (see Fig. 10).

So a' lies at the same time on the 2 conics which already have three common points b', c', d'. So a' is the forth intersection points of these two conics. The explicit solution of this intersection is given in the appendix.

We can thus construct the viewing line Aa'. The same method is used for the other points. Therefore we can reconstruct the optical center O in the scene as the intersection of these lines.

4.3   How to Back Project an Image Point?

Now let's look at how to construct the viewing line of any observed point m in the image plane. We use the same notations as in the previous paragraph.

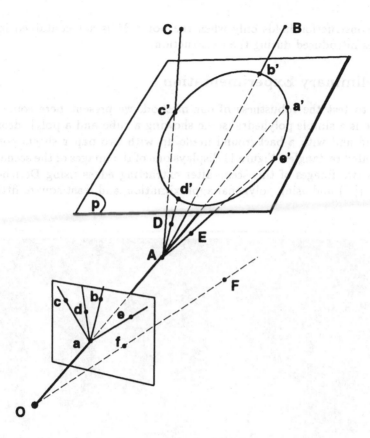

Fig. 10. Determining the view line associated with a known point A

Using the previous reconstruction of the point a' we can determine the pencil of planes going through Aa', so the cross-ratio $[ab, ac, ad, am]$ measured on the image determines the plane OMA in the scene. Similarly, we can take another reference point, for instance B instead of A and determine the plane OBM. These planes intersect in the line OM in the scene which is the viewing line of the point m.

4.4 Where is the Image Point?

Inversely, given a point M in the scene, how to locate it in the image plane?

With O, A, B, C, M as known points, we can construct the line l_a containing a and defined as the intersection of OAM and the image plane, using the cross-ratio $[OAB, OAC, OAD, OAM]$ thanks to:

$$[OAB, OAC, OAD, OAM] = [ab, ac, ad, l_a] \tag{15}$$

Similarly we can get another line, for example l_b. The image point m is then the intersection point of the lines l_a, l_b.

This construction holds only when the point M is not contained in one of the planes introduced during the construction.

5 Preliminary Experimentation

In order to test the robustness of our method, we present here some results. The scene is a simple polyhedral scene showing a cube and a polyhedron in the foreground and with a background made up with two paper sheets containing black painted rectangles. Figure 11 displays one of the images of the scene. Figure ?? shows two images of this scene after extracting edges using Deriche's edge extractor ([5]) and using polygonal approximation, and least square fitting.

Fig. 11. Image of the test scene

Measures of reference points in the scene were performed with a regular ruler, which provides a 1mm accuracy. The two vertical planes were supposed to be orthogonal, in fact their angle is 92 degree.

5.1 Measures in a Plane

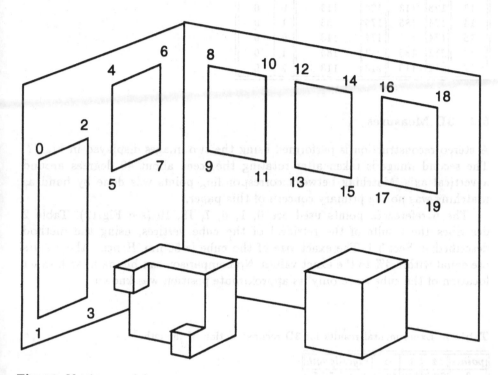

Fig. 12. Vertices used for test

In this first experiment, all the retrieved points lie in a known plane, so we need only one image to derive their 3D position.

We choose four coplanar reference points lying at the vertices of a rectangle; they provide a projective reference frame for the plane wich contains them. We then locate points which are known to belong to this same plane. For this we only need to determine the intersection of each viewing line with the plane, using the method described in Sect.3.1. We then compare the computed locations with the ruler measurements performed directly on the scene.

Table 1 shows the results with points 8, 9, 18, 19 chosen as the reference rectangle.

Using other reference points can lead to very different results. For instance if the points 8, 9, 10 and 11 are taken as reference, points 16, 17, 18 and 19 are measured with errors up to 6mm. This is essentially because these points are far from the reference points and small errors on the reference points are amplified in the measures.

Table 1. Experimental results for planar points

points	real values		computed values		difference	
	x	y	x	y	dx	dy
10	90	185	92.5	185	2.5	0
11	90	113	91	113	1	0
12	128	185	129	185	1	0
13	128	113	128	113	0	0
14	174	185	175	185	1	0
15	174	113	174	113	0	0
16	212	185	213	185	1	0
17	212	113	210	113	2	0

5.2 3D Measures

A stereo reconstruction is performed using the two images displayed in Fig.??. The second image is taken after rotating the scene about 40 degrees around a vertical axis. Matching between corresponding points was done by hand as matching was no the primary concern of this paper.

The 6 references points used are 0, 1, 6, 7, 18, 19 (see Fig.12). Table 2 describes the results of the retrieval of the cube vertices, using the method described in Sect.3.4. The exact size of the cube is 50mm. Hence , the results are equal within 4% to the exact values. No comparison can be made with exact location of the cube since only its approximate position was known.

Table 2. Experimental results for 3D reconstruction of the cube

points	x	y	z	edges	length
0	78.9	140	48.5	0–1	50.5
1	79.1	141	-2	0–2	49.1
2	81.3	189	47.5	0–6	48.9
3	82.0	188	-1.5	1–3	47.1
4	33.2	195	48.5	2–3	49.5
5	34.4	194	-1.5	2–4	48.9
6	30.3	145	49.0	3–5	48.0
				4–5	49.8
				4–6	50.1

However in order to check the values of this reconstruction for precisely located 3D points, we used this technique for locating the same points on the vertical planes. Table 2 provides the results and again the reader can have a guess of the quality.

So, if absolute errors are up to three millimeters, it has to be noticed that differences of such values (relative errors) are also bounded by three millimeters.

Table 3. Experimental results for 3D reconstruction of particular points

points	real values			computed values			difference		
	x	y	z	x	y	z	dx	dy	dz
2	100	0	132	99.5	-1	132	0.5	-1	0
3	100	0	5	100	-0.5	4.5	0	-0.5	-0.5
4	70	0	185	69.5	-0.5	185	-0.5	-0.5	0
5	70	0	113	70	-0.5	113	0	-0.5	0
8	0	44	185	0.5	45.5	185	0.5	1.5	0
9	0	44	113	1	47	114	1	3	1
10	0	90	185	-2	92	185	-2	2	0
11	0	90	113	1	92.5	114	1	2.5	1
12	0	128	185	0.5	130	185	0.5	2	0
13	0	128	113	2	131	114	2	3	1
14	0	174	185	0.5	176	185	0.5	2	0
15	0	174	113	1	176	113	1	2	0
16	0	212	185	0.5	213	185	0.5	1	0
17	0	212	113	0.5	213	114	0.5	1	1

The average value is lower, but this simple experiment had no intend to deliver reliable statistic on these results.

It seems also that systematic errors occur as it can be seen for instance in Table 3 for the y coordinate for points 8 to 15. Such systematic errors may provide from the approximate set up in this preliminary experience. A careful study of stability and error propagation is presently not done and kept for future work.

6 Discussion

6.1 Relative Positioning

Another important aspect of this approach is that it is possible to induce some important relative information with few available structured reference information. This allows for instance to locate objects relatively one to another without knowing the numerical values of their dimensions. As an example we are going to consider the problem of locating a viewing line associated with a point, with respect to a parallelepiped. Choosing such a polyhedral object simplifies the problem by reducing its number of parameters: in Fig.13 the six reference points are entirely defined from the corner A and the vectors $\boldsymbol{AB}, \boldsymbol{AC}, \boldsymbol{AE}$; for instance, the point D is defined by the relation $\boldsymbol{AD} = \boldsymbol{AB} + \boldsymbol{AC}$.

For this purpose, we need first to be able to locate a point from its projection coordinates, having a parallelogram as a reference. For instance, we compute the position of the viewing line associated with the point M in the parallelogram $ABDC$ (see Fig.13).

Let:

$$k_1 = [AB, AC, AD, AM] \tag{16}$$

157

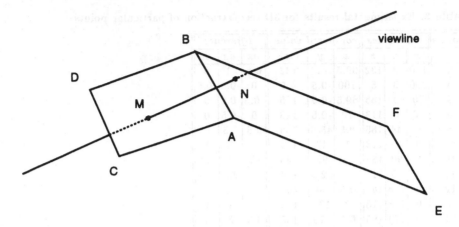

Fig. 13. Locating a viewing line toward a parallelepiped

$$k_2 = [BA, BD, BC, BM] \tag{17}$$

where k_1 and k_2 have been computed from the image points.

One can show that in the case of $ABDC$ being a parallelogram,

- the vector $\boldsymbol{AM}$ is colinear to $\boldsymbol{AC} + k_1 \boldsymbol{AB}$
- the vector $\boldsymbol{BM}$ is colinear to $\boldsymbol{AC} - k_2 \boldsymbol{AB}$

This determines the two lines AM and BM and thus their intersection, M. The position of M can be expressed with respect to the basis of the parallelogram, $(A, \boldsymbol{AB}, \boldsymbol{AC})$ and the cross-ratios k_1 and k_2:

$$\boldsymbol{AM} = \frac{1}{k_1 + k_2}(\boldsymbol{AC} - k_1 \boldsymbol{AB}) \tag{18}$$

In the same way, the other point of the viewing line, N can be expressed in the parallelogram $ABFE$ as a function of the cross-ratios:

$$k_3 = [AB, AE, AF, AN] \tag{19}$$

$$k_4 = [BA, BF, BE, BN] \tag{20}$$

A parametric equation of the viewing line NM can then be given in the reference system of the parallelepiped referred $(A, \boldsymbol{AC}, \boldsymbol{AE}, \boldsymbol{AB})$: $P \in NM \Leftrightarrow \exists \lambda$:

$$\boldsymbol{AP} = \frac{1-\lambda}{k_1 + k_2}\boldsymbol{AC} + \frac{\lambda}{k_3 + k_4}\boldsymbol{AE} - \left[\frac{(1-\lambda)k_1}{k_1 + k_2} + \frac{\lambda k_3}{k_3 + k_4}\right]\boldsymbol{AB} \tag{21}$$

Notice that the viewing line is completely defined by the four parameters k_1, k_2, k_3 and k_4. Lines in the 3D space passing through a point O have two degrees of freedom, therefore these four parameters are linked together by two relations (see Appendix).

6.2 Euclidean Properties

Although we can fulfill the same fundamental visual tasks as the classical calibration does, we lack a method to represent some usual Euclidean metric information such as rotation. The projective properties used through-out this paper totally hide transformation like rotation. So *how to make it explicit?* That is, *how and when to come back to the usual space*, and *what kind of right tools should be used to do this?*

Certainly, partial answers could be found by introducing the circular points, as they define the metric of the affine plane embedded in the projective plane. Thus Laguerre's formula provides us with a natural way to represent angle metrics. We are working in this direction for the time being.

Another interesting direction is the use of points at infinity, well-known as vanishing points in the computer vision community [11]. These points at infinity represent the directions of Euclidean lines. They have been widely used to recover spatial line directions, thus providing the rotational information. How to use it more systematically and relate it to the remains of the paper is still an open problem.

When the camera is perfectly calibrated, the tools developed here are too general. In this case more Euclidean properties can be obtained directly from the projections. Partial answers can be found in the photogrammetry literature [15].

6.3 How to Apply it?

One easy application of these technics is to have a reference frame within the scene. This is possible for controlled environment like for some of the industrial or medical applications. It may be particularly useful when the working condition does not allow careful calibration. We found it particularly useful because our image grabbing system was unstable, so no calibration was possible.

For systems where careful calibration is impossible, and for which the image is stable, the location of image reference points can be kept in memory. If the whole system moves the coordinate of these references points can be considered as the image of the reference points moving with the overall system. In this case, the storage of the reference point images can be viewed as an easy calibration which allows camera centered vision.

Applications to relative positioning can be done as showed in Sect.6.1. This point needs however much further exploration. In particular even with simple expression of the results, it is difficult to take into account large uncertainty on the input data, like for example: the heights of the house is in-between 4m and 8m.

7 Conclusion

We have shown that the fundamental problems in computer vision, such as

- how to back project an image point?
- how to project a space point?
- how to determine epipolar geometry?

can be resolved by direct geometric construction with known reference points using strictly projective properties.

Therefore,

- No direct tedious numerical calibration is needed, only reference points are necessary.
- The method involves only simple computation, almost always arithmetic computation of cross-ratio.
- The method does no less work than traditional calibration, but more in some special cases, for example, geometric inference from partial reference information.
- Only pure projective invariant properties are used.
- The method gives an intuitive display of calibration process, instability phenomena can be easly singled out.

An experimental study has shown that the method is reliable as long as the measures of the reference points are reliable. Its accuracy gracefully degenerates when the measures are done on points located far from the reference points. Study has still to be done in order to integrate the behavior of the whole process when dealing with inaccuracy. The use of redundant data in order to improve the accuracy is an other way to continue the work presented here.

Appendix

Intersection of two Conics

We provide here the explicit solution of the intersection of two conics having three common points. The conics are defined as in Sect.4.1 and the three common points A, B and C are respectively the origin and extremities of both unit vectors on the axis. Such a simple reference frame change is always possible and it considerably simplifies the complex formula we obtain.

The first conic C_λ is defined by the cross-ratio λ and its forth point $D = (x_d, y_d)$ (see Sect.4.1). The second conic C_μ is defined by the cross-ratio μ and its forth point $F = (x_f, y_f)$.

Using symbolic computation tool (we used Maple) we derive x and y, coordinates of the forth intersection point of C_λ and C_μ.

$$x = \frac{Z(\mu x_f(1 - y_d - x_d) - \lambda x_d(1 - x_f - y_f))}{\Delta}$$

$$y = \frac{Z(\mu y_f(1 - y_d - x_d) - \lambda y_d(1 - x_f - y_f) + x_d y_f - y_d x_f - y_f + y_d)}{\Delta}$$

where

$$Z = \mu x_f y_d (\lambda - 1) + \lambda x_d y_f (1 - \mu)$$
$$\Delta = \quad \lambda\mu\,[-x_f(1 - y_d) + (1 - y_f)x_d]$$
$$[\lambda(\mu - 1)y_f(1 - y_d) - y_d(\lambda - 1)(1 - y_f)\mu$$
$$+ 2(\mu - 1)(\lambda - 1)(y_d(1 - x_f) - (1 - x_d)y_f)]$$
$$- (\mu - 1)(\lambda - 1)\,[y_d(1 - x_f) - (1 - x_d)y_f]$$
$$[\lambda(\mu - 1)x_d(1 - x_f) - \mu(\lambda - 1)(1 - x_d)x_f]$$
$$+ [\lambda(\mu - 1)y_f x_d - \mu(\lambda - 1)x_f y_d]$$
$$[\lambda(\mu - 1)(y_f x_d - (1 - x_f)(1 - y_d))$$
$$+ \mu(\lambda - 1)((1 - y_f)(1 - x_d) - x_f y_d)]$$

Link Between the Cross-ratios

In Sect.6.1 the equation of the viewing line associated with an image point is defined with respect to a reference parallelepiped (A, B, C, D, E, F) and as a function of the cross-ratios k_1, k_2, k_3 and k_4, which have been determined from measures in the image.

Since lines in the 3D space passing through a point O have two degrees of freedom —for instance the two angles which define its direction— the four parameters k_i are to be linked together by two relations. If we express that the optical center O is the intersection of the viewing lines associated with C and D and that any viewing line passes through O, we get two relations linking respectively k_1 and k_3, k_2 and k_4 :

$$k_1 = \frac{l_{3c} - k_3}{l_{3d} - l_{3c}}$$

$$k_2 = 1 - \frac{l_{4c} - k_4}{l_{4d} - l_{4c}}$$

where :

$$l_{3c} = [AB, AE, AF, AC']$$

$$l_{4c} = [BA, BF, BE, BC']$$

$$l_{3d} = [AB, AE, AF, AD']$$

$$l_{4d} = [BA, BF, BE, BD']$$

with C' and D' intersections of OC and OD with the plane $ABFE$.

161

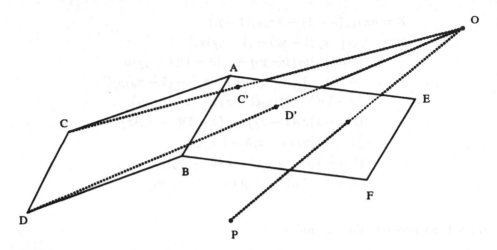

Fig. 14. The constraint that all the viewing lines must contain O leads to relations linking the parameters of the equation of a viewing line.

References

1. N. Ayache. *Stereovision and sensor fusion*. MIT-Press, 1990.
2. N. Ayache and F. Lustman. Fast and reliable passive strereovision using three cameras. In *Proceedings of International Worshop on Machine Vision and Machine Intelligence, Tokyo, Japan*, 1987.
3. D.H. Ballard and C.M. Brown. *Computer Vision*. Prentice Hall, 1982.
4. J.B. Burns, R. Weiss, and E.M. Riseman. View variation of point set and line segment features. In *Proceedings of DARPA Image Understanding Workshop, Pittsburgh, Pennsylvania, USA*, pages 650–659, 1990.
5. R. Deriche. Using Canny's criteria to derive a recursively implemented optimal edge detector. *International Journal of Computer Vision*, 1(2):167–187, 1987.
6. R. Duda and P.Hart. *Pattern Classification and Scene Analysis*. Wiley-Interscience, 1973.
7. N. Efimov. *Advanced Geometry*. Mir, Moscow, 1978.
8. O.D. Faugeras and G. Toscani. Camera calibration for 3D computer vision. In *Proceedings of International Workshop on Machine Vision and Machine Intelligence, Tokyo, Japan*, 1987.
9. R. Horaud, B. Conio, O. Leboulleux, and B. Lacolle. An analytic solution for the perspective 4-point problem. *Computer Vision, Graphics and Image Processing*, 47:33–44, 1989.
10. S.J. Maybank. The projective geometry of ambigious surfaces. Technical Report 1623, Long Range Laboratory, GEC, Wembley, Middlessex, UK, July 1990.
11. L. Quan and R. Mohr. Determining perspective structures using hierarchial Hough transform. *Pattern Recognition Letters*, 9(4):279–286, 1989.
12. J.G. Semple and G.T. Kneebone. *Algebraic Projective Geometry*. Oxford Science Publication, 1952.

13. G. Sparr and L. Nielsen. Shape and mutual cross-ratios with applications to the interior, exterior and relative orientation. In *Proceedings of the 1st European Conference on Computer Vision, Antibes, France*, pages 607–609. Springer-Verlag, April 1990.

14. C. Trip. Where is the camera ? *Mathematical Gazette*, 71:8–14, 1987.

15. P.R. Wolf. *Elements of Photogrammetry*. McGraw-Hill, New-York, 1974.

16. Y. Liu and T.S. Huang and O.D. Faugeras. Determination of camera location from 2D to 3D line and point. IEEE *Transactions on PAMI*, 12(1):28–37, January 1990.

17. J.S.C. Yuan. A general phogrammetric solution for the determining object position and orientation. IEEE *trans. on Robotics and Automation*, 5(2):129–142, 1989.

Geometrical Representation of Shapes and Objects for Visual Perception

Jean-Marc Chassery and Annick Montanvert

Equipe RFMQ-TIMC, CERMO BP 53X
38041 Grenoble cedex France

Abstract. When visual perception is modelized on computer, some geometrical formalisms are present at different levels of the approach. This paper deals more precisely with the choice of an adapted representation for image analysis shape understanding. In this context, we present several geometrical models for representation of images and shapes useful for their description and interpretation.

First we present geometrical models to split an image, then geometrical models for shape representation and shape description. The representation space, that is discrete space or euclidean space, is pointed out. These methods also depend on the implementation of data structures and algorithms. Finally we present the three-dimensional extensions of the two-dimensional models.

1 Introduction

This paper illustrates optimal methods for shape representation defined in order to facilitate their processing and understanding. These methods will involve structural approaches in opposition to statistical approaches.

Methods for coding will be distinguished from methods for representation. The coding methods are mainly devoted to data compression and they are evaluated in terms of compression ratio and reversibility ratio. The representation methods are devoted to the description of coded objects and they are selected on their ability to extract some shape characteristics or to speed-up some processes.

In the first class we will find for instance run-length encoding, which is a very poor approach on the structural point of view. In the second class, we will find medial axes which integrate some description properties.

The sampling process attached to the acquisition phase provides a spatial representation of an image associated with a discrete referential of $\mathbb{Z}^2$. Such a coding is only a support of information and is very often inadequate to study a phenomenon in the image. It should also be useful to take advantage of the discrete nature of this representation.

The definition of shape understanding methods in the context of the discrete space involves to pay attention to the algorithmic developments. Some constraints on algorithmic schema and on data structures are necessary. Such developments are in close relation with computer architectures.

This paper begins by a presentation of models for tiling the plane, followed by representation models and descriptive models. The strong connection between Euclidean models and discrete models when working on binary images will be shown.

2 Image and geometrical representation

2.1 Models for tiling

Geometrical tessellations of the plane (see Figure 1) begin with the famous Escher's drawings, in which complex and judicious imbrications of patterns have been imagined. Simpler are the three regular tilings extracted for the eleven Archimedean tilings and defined by the regular polygons which are the square, the triangle and the hexagon [1]. In image analysis these three regular tilings are used, and also geometrical tessellations defined by Voronoi diagrams.

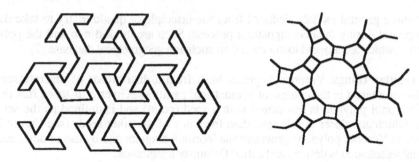

Fig. 1. Examples of geometrical tilings of the plane.

Regular tilings. Regular tilings are the tessellations defined by the same regular polygon all over the plane, and such that each vertex is adjacent only with other vertices (edges are forbidden).

Among the three regular tilings, the square tessellation is built with patterns sharing a unique orientation. Moreover, it has the property of being recursive. This last property is at the origin of the quadtree representation (see Figure 2).

A quadtree is defined by a recursive decomposition of an image into quadrants, until the blocks are homogeneous. For a 2^n x 2^n image, the size of a block at level i is 2^i x 2^i, and this block is allowed to take 2^{n-i} x 2^{n-i} different positions [2]. Thus the quadtree coding is closely linked to the discrete space.

Because of its structure which is a quaternary tree, a quadtree can be easily encoded; access to pixels and to their neighbors are equivalent to displacements inside the quadtree. The process to build the quadtree can be very efficient, when a scanning order of the pixels in the original image, such as the Z-Peano order is used. Then each pixel is scanned only once, and the nodes are created only when they must remain in the final quadtree.

The quadtree representation is efficient for transformations on sets or geometrical operations. But the constraints on the location of the blocks make it very sensitive to translations, which is not acceptable for shape description. The Quadtree Medial Axis is an incomplete solution to this problem.

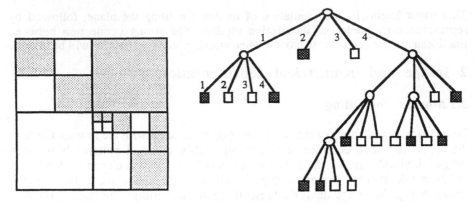

Fig. 2. An image and its quadtree.

Some more general models, deduced from the principle of quadtree, try to take data into account during the decomposition process. Such are the k-d tree and the point-quadtree, which are adapted to punctual data such as a geographical database [2].

Non regular tilings. Voronoi diagrams built from a distribution of discrete seeds are also of interest in the context of geometrical partitions. Given a distance function, each Voronoi polygon is associated with a seed (point) and is defined as the set of points which are closer to this seed than from any of the other seeds (see Figure 3). The set of Voronoi polygons generates the Voronoi diagram and by use of adjacency, the dual tessellation is defined and called Delaunay tessellation.

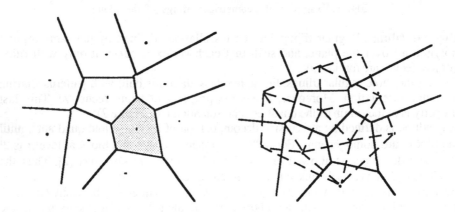

Fig. 3. A Voronoi diagram and the associated Delaunay tessellation.

Using the euclidean metric, the generation model of Voronoi diagram is based on a dedicated data structure. Such a structure allows the computation of mediators between adjacent seeds in order to generate the boundaries of polygons.

Algorithmic developments for Voronoi diagram computation can be split into two classes: Divide and Conquer approach and Incremental approach. Divide and Conquer approach has been proposed by Shamos and Hoy in 1975 and Lee in 1976 [3]. It is a recursive method with a complexity in O(n.log n), where n represents the number of seeds. The main problem of this approach is that it needs to recalculate the whole diagram when only a few seeds are changed.

The incremental method is iterative. Such a method allows to add or to remove some seeds without recalculating the whole diagram, because the computation is performed locally. Such a method has been proposed by Green-Sibson [4] and enhanced by Ohya-Iri-Murota [5].
Each of these two methods provides the resulting partitionning with an adapted graph structure in which access to adjacency between polygons is easy; it also allows the use of mathematical morphology on graphs [6].

As for quadtrees such a model can be extended to other types of data. For example Voronoi diagrams have been computed with non ponctual seeds such as straight lines or polygons. This defines the Generalized Voronoi diagram and its extension to SKIZ (SKeleton by Influence Zone) (see Figure 4) [7, 8].

Fig. 4. Generalized Voronoi diagram and SKIZ.

In Section 3.2, we will present the definition of Voronoi diagrams in discrete space using discrete distance functions.

2.2 On the use of tilings

The most important illustrations on the use of such geometrical tilings are image coding and image compression.
The quadtree shows a hierarchical organization based on the resolution in an image. Such an organization allows to define some filtering processes on shapes (see Figure 5). Consequently we can speak of coding by "depth". A mean compression ratio can be evaluated [9]. A main advantage versus the matrix coding is that it is resolution-invariant.
The data structure related to Voronoi diagram is a graph allowing access to adjacency between regions. It does not provide a hierarchy like with quadtree model but it allows

the access to an organization between primitives. Voronoi diagrams have been used for image segmentation [10]. The method consists in defining a split and merge process initialized from a random set of seeds on the image support.

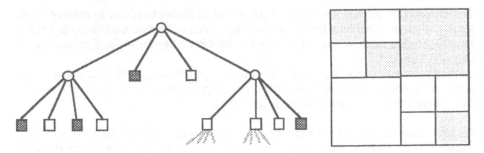

Fig. 5. Approximation by the interior on a quadtree representation.

During the split step, new seeds are added inside polygons which are not homogeneous (criterion based on statistical orders such as standard deviation or max-min deviation). The merge step allows the fusion of adjacent polygons by the use of a similarity criterion (criterion based on mean values). Such an iterative approach is illustrated in Figure 6.

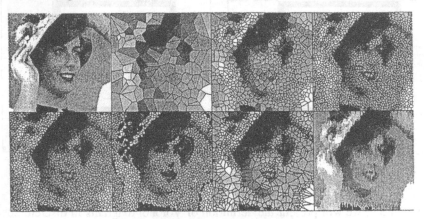

Fig. 6. Segmentation using a Voronoi diagram structure.

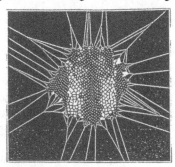

Fig. 7. Voronoi partitionning and an interpretation by similarity.

The exploitation of SKIZ is usual in biomedical applications in order to assign some influence zones to various entities as it is illustrated in Figure 7 [11].

3 Shape and geometrical models of representation

A digital image is strongly associated with its discrete support, which is the square lattice in most of the cases. But the objects in an image (i.e. connected components) are localized at some random positions. Consequently, in order to improve the quality of a description, it is fundamental to take into account the objects rather than the lattice to define a model of representation.

A shape is usually described by its contour or by the associated connected component of pixels [12]. Thus we will successively describe contour based models and region based models.

3.1 Contour based representation models

When considering the pixels, the contour is defined as a sequence of boundary points. Such a sequence is represented and analyzed with the Freeman encoding. A syntactic analysis of this sequence provides the extremities of digital straight lines; we speak of vectorization process. Such a stage is reversible because the contour can be drawn again by the use of a discretization method such as the Bresenham's algorithm. With this approach we deal with the discrete space only.
Using iterative or recursive approximation methods, a polygonal approximation of the contour can be extracted; the resulting straight lines are issued from euclidean geometry. Nevertheless the loss of information is balanced by an increasing of the compression ratio of information in the image.

Generalized Voronoi. The data structure used in a polygonal approximation includes the extremities of the straight lines and is an implicit way to access to the sides of the polygons. Starting from this data structure, a partition of the shape delimited by the contour can be generated. For that, to each element involved in the contour coding, the set of pixels closer to this element than from the others is associated (see Figure 8).

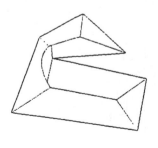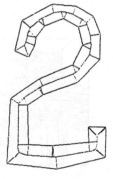

Fig. 8. Shape partitionning based on the Generalized Voronoi diagram.

Such an affectation is provided by computation of the Generalized Voronoi diagram in which the seeds are extended to the extremities and the corresponding open straight lines. The computation of the Generalized Voronoi diagram uses a Divide and Conquer approach for the interior of the shape [8].

If the influence zones associated with concave extremities are removed, the simplified diagram is the skeleton of the shape (see Figure 9).

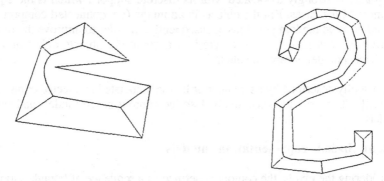

Fig. 9. Skeleton deduced from the Generalized Voronoi diagram.

Generalized Voronoi diagram and discretization. A second approach to realize the shape partitionning from its contour consists in computing a regular discretization of the contour and in associating to each ponctual seed its Voronoi polygon. Such a partition involves the algorithm based on pixels as seeds. In a second step polygons corresponding to a same initial straight line are grouped (see Figure 10).

Using this approach, a discrete approximation of the skeleton is computed. This approach has been used in a region/contour cooperation environment for segmentation [13].

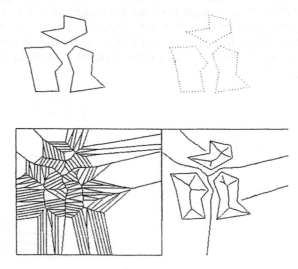

Fig. 10. Partitionning and skeleton using Voronoi diagram based on pixels.

Discrete propagation. The discretization can be totally used, when computing the euclidean distance of pixels in the foreground to the background. The two images providing the displacements in abscissa and ordinate to reach the closest point on the contour are calculated [14], which provides the partition from pixels as seeds. This approach is close to discrete distances (see Section 3.2).

3.2 Skeleton representation

Until now, we have presented partitionning methods. On another side, there exist some methods based on shape covering issued from discrete distances, which provide medial axes and medial lines for representation, description and analysis of shape. They come from the skeleton representation, which was firstly defined in euclidean space and that we already evocated when presenting Generalized Voronoi diagrams.

Continuous description. The first studies on skeletons for shape description started in the sixties, with the definition of skeletons in the continuous space [15]. A skeleton is a *thin, centered and reversible representation* of a shape, which gives a good idea of the shape of the original object. It is sometimes called symmetry-axis. Its interest is well illustrated when analysing thin objects, such as characters or chromosomes. Moreover it is also useful for thick objects when the distance function to the background is added on the skeleton: then the skeleton can be considered as an oriented planar graph (see Figure 11).

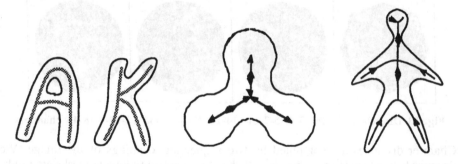

Fig. 11. Examples of shapes and their skeletons.

The process is usually illustrated with the propagation of a wave initialized on the boundary of the shape: the loci where the wavefront extinguishes itself define the skeleton, and the associated dates provide the distance function.

To compute the skeleton in continuous space, several methods have been developed, for instance analytical simulation of the wavefront, or detection of symmetries, or extraction of local maxima of the distance function. The first approach is related to generalized Voronoi diagrams computed from polygonal approximation of the boundary; the skeleton is calculated as composed of straight lines and arcs of parabola [16, 17].

The main difficulty of skeletonization is to define efficient algorithms in discrete space. Then two classes of methods can be distinguished:

- methods based on iterative removing of contour pixels; they provide binary skeketons;

- methods based on discrete distance function; they provide medial axes and medial lines.

Thinning processes. Let S be the set of foreground pixels; SK is a binary skeleton of S if:
- it is included and centered in S,
- it is thin, that is composed of arcs and curves,
- it shares with S the same number of connected components and holes (the transformation is said to be homotopic),
- it has some branches where the object has some elongated parts,
- if some points are removed from SK, one of the two previous properties will be respected anymore.
Some of these properties are rather subjective, and they do not define any computation algorithm directly.

The processes which are usually implemented are parallel and work on the contour pixels. At each iteration, simple points (pixels which are not necessary to maintain the connectivity of the remaining set of pixels) which are not terminal points (not significant of an elongated subpart of the shape) are removed [18]; these pixels are detected by local tests. Numerous variants of thinning algorithms have been proposed in the literature, and they provide slightly different skeletons for a given shape [19].
Unfortunately a binary skeleton is not reversible. A partial reversibility could be provided by superposing a skeleton with a discrete distance computed on the initial shape. But in that case, medial lines are prefered (see Section 4).

Discrete distances. These distances are related to the discrete space, and they induce the second transcription of continuous skeletons, which are medial axes.

The basic distances on square lattice come from 4- and 8-adjacency:
$d_4 (P, Q) = |i_P - i_Q| + |j_P - j_Q|$ *City Block Distance* (ou *Square Distance*).
$d_8 (P, Q) = \max (|i_P - i_Q| , |j_P - j_Q|)$ *Chessboard Distance* (ou *Diamond Distance*).
But they are far away from the values of the euclicean distance, because the discrete disks that they generate are squares.

The euclidean distance can be computed, but two matrices or floating numbers must be manipulated. The associated algorithms are parallel [20] or sequential [14], or they use some special data structures [21, 22] but they are not as simple as for chamfer distances that we present below, and which complexity is in $O(n^2)$.
Hopefully, discrete distances can be defined, which are closer in value to the Euclidean distance, and which are easily and efficiently computed on a sequential computer; the result is a distance image (distance map), in which each pixel is assigned with an integer value.
Chamfer distances are based on a weighting of the elementary displacements on the lattice, in order to minimize the error when compared to the euclidean distance.
Neighborhoods of different sizes can be used: the larger the neighborhood, the better the approximation.
For the 3x3 neighborhood, the distance called 3-4 chamfer is well known: the direct displacement is weighted by 3, and the indirect displacement is weighted by 4. Then the maximum relative error compared to the euclidean distance is equal to 5,72%,

rather than 29,29% for the chessboard distance d$_8$. The weights are calculated in order to minimize this relative error on a square image [23].

When increasing the size of the neighborhood, 5-7-11 chamfer for 5x5 neighborhood (1,98% of relative error) and 14-20-31-44 chamfer for 7x7 neighborhood (1,52% of relative error) are defined (see Figure 12).

$$
\begin{pmatrix} 4 & 3 & 4 \\ 3 & 0 & 3 \\ 4 & 3 & 4 \end{pmatrix}
\qquad
\begin{pmatrix}
2x7 & 11 & 2x5 & 11 & 2x7 \\
11 & 7 & 5 & 7 & 11 \\
2x5 & 5 & 0 & 5 & 2x5 \\
11 & 7 & 5 & 7 & 11 \\
2x7 & 11 & 2x5 & 11 & 2x7
\end{pmatrix}
$$

Fig. 12. Weights in 3x3 and 5x5 neighborhoods.

When minimizing directly in the discrete space, the quality of the distance can still be improved, and the relative error reduced with 12-17-°-38-43 chamfer in a 7x7 neighborhood (1,38% relative error) (see Figure 13) [24]. The directions of the elementary displacements are defined by the Farey series, and some properties of these distance functions can be deduced [25].

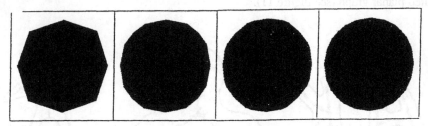

Fig. 13. Discrete disks for 3-4, 5-7-11, 14-20-31-44-° and 12-17-°-38-43 chamfers.

Chamfer distances are computed by two sequential scannings of the image. We describe below the algorithm for the 3-4 chamfer. The two following masks are used:

$$
m_{P\text{-forward}} = \begin{pmatrix} 4 & 3 & 4 \\ 3 & 0 & \end{pmatrix}
\quad \text{and} \quad
m_{P\text{-backward}} = \begin{pmatrix} & 0 & 3 \\ 4 & 3 & 4 \end{pmatrix}
$$

Then the following algorithm is applied on the image:

Forward scanning:
for i = 1 to N do
        for j = 1 to M do
                A(i,j) = minimum $_{(k,l)\ \in m_{P\text{-forward}}}$ (A (i+k, j+l) + m$_{P\text{-forward}}$ (k,l))
Backward scanning:
for i = N to 1 do
        for j = M to 1 do
                A(i,j) = minimum $_{(k,l)\ \in m_{P\text{-backward}}}$ (A (i+k, j+l) + m$_{P\text{-backward}}$ (k,l))

The improvment of the distance images is obvious when the distances are computed from pixels, as on a Voronoi diagram (see Figure 14).

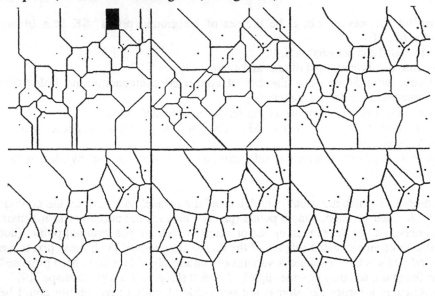

Fig. 14. Distance images for discrete distances: d_4, d_8, d_{octog}, , $d_{3\text{-}4}$, $d_{5\text{-}7\text{-}11}$, d_E.

4 Shape description

4.1 Medial axis

Based on discrete distances, medial axes are the discrete transcription of the characterization of skeletons involving a covering of a shape by maximal digital disks.

A digital disk centered on P, with a radius r is: $B_d(P, r) = \{Q : d(P,Q) \le r\}$
Let S be a shape, the background is noted F.
For all P in S, $B_d(P, R_p)$ where $R_p = d(P, F) - 1$ is included in S.

Therefore, $S = \cup_{P \in S} B_d(P, R_p)$.

The medial axis, noted AM(S), is characterized by [26, 27]:

$$P \in AM(S) \Leftrightarrow \forall\ Q \in S, B_d(P, R_p) \not\subset B_d(Q, R_Q)$$

The medial axis is defined by the centers of the maximal disks, and their associated radii. For each point of AM(S), the associated disk is not totally included in any other disk. The shape is covered by the maximum discrete disks (note that a pixel might belong to several maximum disks). From the previous formula, it is obvious that S is completely defined by its medial axis AM(S), and then:

$$S = \cup_{P \in AM(S)}\ B_d(P, R_P)$$

The first models of medial axes were deduced from d_4 and d_8 distances (see Figure 15).

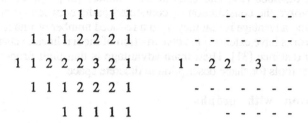

```
          1 1 1 1 1              - - - - -
        1 1 1 2 2 2 1          - - - - - - -
    1 1 2 2 2 3 2 1        1 - 2 2 - 3 - -
        1 1 1 2 2 2 1          - - - - - -
          1 1 1 1 1              - - - - -
```

Fig. 15. Medial axis extracted from a distance image for d_8.

The medial axis points of an object S are identified by:
$$P \in AM(S) \quad \Leftrightarrow \quad \forall \ Q \text{ neighbor of } P, \ R_P \geq R_Q$$

This is performed during one scanning of the distance image, and then the process is very efficient. The reconstruction of a shape from its medial axis is also very easy to implement, using a process similar to the distance function computation.
Computing medial axes for other chamfer distances is not that easy [28].

Because of the chessboard distance, the discrete disks are squares, and then they can be compared to quadtrees:
- a quadtree is a decomposition of a shape into homogeneuous squares which location is constrainted by the lattice. Then it is easy to encode, but not very convenient for shape description.
- a medial axis covers a shape by homogeneous squares, which may partially overlap, and location of which is defined by the shape itself. This representation is shift-invariant, and then each square is significant of the shape.

4.2 Medial line representation

Medial axis are disconnected, centered, sometimes locally thick, reversible. They are characterized by the use of a distance transform. In order to provide shape understanding processes, some relations must be defined between medial axis centers. Medial lines will offer these two main advantages: reversibility and description power.

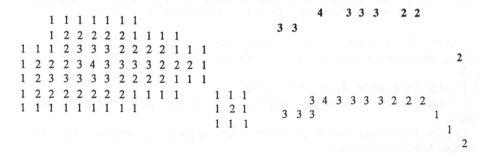

Fig. 16. An object, its medial axis and its medial line for the chessboard distance.

The construction of medial line from medial axis is done by specific algorithms based on city block distance [29] and chessboard distance [30] (see Figure 16). These algorithms involve the construction of connected paths between components of the medial axis. One advantage is that they need a reduced number of image scannings.
More recent works have extended the construction of medial lines to chamfer distances and euclidean distance [31]. Their main advantage is that their property of isotropy provides robust tools for shape description in discrete space.

4.3 Description with graphs

Medial axis and medial line are close to the continuous model of shape and thus close to descriptive properties. Points of medial axis are put in relation by medial line. In fact, each point of the medial axis gives a new contribution to the object formation in comparison with the other points of the medial axis. An analysis of the structure of the object is performed by a traversal of the medial line and many works are focused on filtering or decomposition processes.

The medial line can be decomposed into arcs, each of them defines a subpart of the shape [32] (see Figure 17), for that some decomposition rules must be defined.

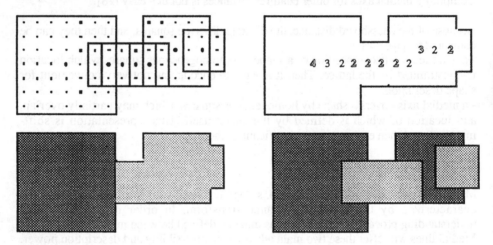

Fig. 17. Medial axis and covering by squares. Connection with medial line. Exemples of shape partitionning induced by medial line decompositions.

The medial line contains both the topological and hierarchical information about the shape, and it can be expressed more formally using a graph, called medial line graph [33].
It is a directed graph in which (see Figure 18):
- the nodes are the connected components of medial axis points sharing the same value;
- the arcs are the discrete connection paths, directed towards increasing values of the distance transformed picture.
Each node is an encoding of an elementary surface of the shape (via the medial axis points), the arcs show the hierarchy between these surfaces.

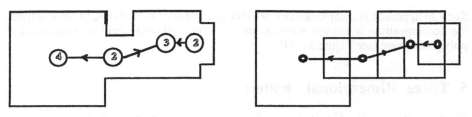

Fig. 18. Medial line graph and organization of shape into subparts.

Modification on a shape can now be deduced from traversals of its medial line graph, sub-graphs extractions and application of the reverse transformation to reconstruct the modified shape [34] (see Figure 19).

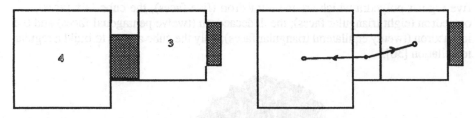

Fig. 19. Shape filtering from: medial axis; medial line graph with connectivity control.

An important application is shape decomposition. It is widely illustrated in separation of aggregates (see Figure 20) and in partitionning of complex objects [35].

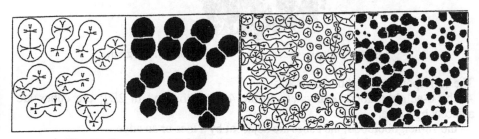

Fig. 20. Separation of aggregates by analysis of medial line graph.

Fig. 21. Decomposition in convex elements of a polygonal shape.

Such an approach is complementary of other decomposition methods, as for example the decomposition in convex elements using Voronoi diagram built on vertices of a polygonal shape (see Figure 21) [4].

5 Three dimensional models

Similar to the matrix of pixels in two dimensional space, the basic three dimensional representation is a three-dimensional matrix of voxels (volume element).

5.1 Partitionning

The geometrical tessellation of a volume is performed using polyedra. Among the five regular polyedra which are the tetraedron (five faces), the cube (six faces), the octoedron (eight triangular faces), the dodecaedron (twelve pentagonal faces) and the icosaedron (twenty equilateral triangular faces), only the cube allows to build a regular tessellation [36].

Fig. 22. Volume decomposed by 3D Voronoi diagram.

The corresponding neighborhoods are:
— the 6-neighborhood (equivalent to the 4-neighborhood in 2D); the cubes are adjacent by a face;
— the 18-neighborhood; the cubes are adjacent by a face or an edge;
— the 24-neighborhood (equivalent to the 8-neighborhood in 2D); the cubes are adjacent by a face or an edge or a vertex.
Because of the amount of data that is implicated by the 3D space, the data structures and the algorithms are of importance.

The quadtree principle is extended to octree: the recursive process splits a cube into eight cubes of half side [37].
As for 2D images, we can define non regular models and for example generalize the 2D Voronoi diagrams in 3D Voronoi diagrams (see Figure 22) [38, 39].

5.2 Skeletons and medial lines

The idea of binary skeleton is also extended to 3D space. Nevertheless, it is more difficult to develop the methods than in the 2D space, mainly because notions such as terminal points or thinness are not so easy to characterize. We can speak of thinness along one direction of the space, which defines surface skeletons. Or we can speak of thinness along two directions of the space, which provides wireframe skeletons. These two concepts are illustrated in Figure 23 [40].

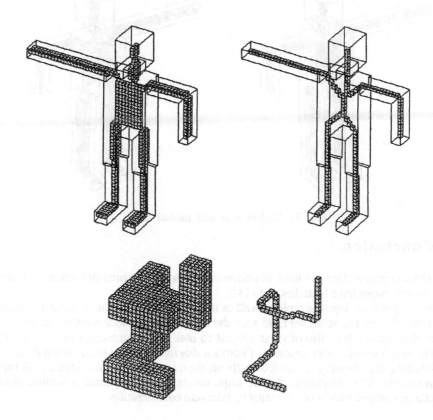

Fig. 23. Surface and wireframe skeleton of a 3D object ; a 3D object and its skeleton.

The chamfer distances can be computed with the same sequential schema than for 2D space, extended to 3D space [41]. We show in Figure 24 the extension of 3-4 chamfer which is 3-4-5 chamfer.

Forward scanning

5	4	5
4	3	4
5	4	5

4	3	4
3	0	

Backward scanning

	0	3
4	3	4

5	4	5
4	3	4
5	4	5

Fig. 24. Forward and backward scannings for 3-4-5 chamfer.

The extraction of medial axes and medial lines have also been developed in 3D [40].

Figure 25 illustrates the medial axis and the medial line computed with the metric d_{26}.

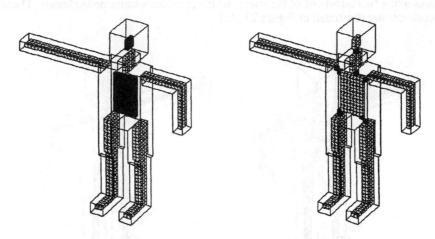

Fig. 25. Medial axis and medial line.

6 Conclusion

The most important two and three dimensional geometrical representation models for images and shapes have been described [34].
Geometry plays an important role as well as in euclidean space than in discrete space. All along this presentation, we have seen the adequation of models either for coding or for description. It is also of great interest to note that continuous models need to prepare some specific data structures (Voronoi diagrams for instance), when discrete models have the advantage to work directly on the discrete image (distance transforms for instance). The constraints for the implementation (algorithmic schemas, data structures, computer architecture) must be take into consideration.

Acknowledgements. Figures 1, 2, 3, 5, 6, 11, 16, 17, 18, 19 have been reproduced from [34] with the authorization of Hermès editor.

References

1.   B. Grunbaum, G.C. Shepard:Tilings and patterns : an introduction. Freeman & co ed. (1989)

2.   H. Samet: Applications of spatial data structures. Addison Wesley (1990)

3.   F.P. Preparata, M.I. Shamos: Computational geometry, an introduction texts and monographs in Computer Science. Elsevier Science Publishers, North Holland (1987)

4.   B. Chazelle, D.P. Dobkin: Optimal convex decomposition. In Computational Geometry, Toussaint G.T. ed, Elsevier Science Publ. (1985)

5. T. Ohya, M. Iri, K. Murota: A fast Voronoi diagram algorithm with quaternary tree bucketing. Information Processing Letters 18, 227-231 (1984)

6. L. Vincent: Mathematical morphology on Graphs. Signal Processing 16(4), 365-388 (1989)

7. J. Serra: Image Analysis and Mathematical Morphology, part II: theoretical advances. Academic Press (1988)

8. H.T. Hu, J-M. Chassery: Squelette et diagramme de Voronoï généralisé en compréhension de formes. Conf. Reconnaissance des Formes et Intelligence Artificielle, AFCET-INRIA, Lyon, pp. 859-864 (1991)

9. C.R. Dyer: The space efficiency of quadtrees. Computer Vision Graphics and Image Processing 19, 335-348 (1982)

10. J-M. Chassery, M. Melkemi: Diagramme de Voronoï appliqué à la segmentation d'images et à la détection d'événements en imagerie multi-sources. Traitement du Signal 8 (3), 155-164 (1991)

11. E. Bertin, F. Parazza, J-M. Chassery: Segmentation and measurements based on 3D Voronoi diagrams: application to confocal microscopy. Computerized Medical Imaging and Graphics, in press (1992)

12. T. Pavlidis: Structural Pattern Recognition. Springer Verlag (1977)

13. M. Melkemi, J.M. Chassery: Edge-region segmentation process based on generalized Voronoi diagram. 11th Int. Conf. on Pattern Recognition, pp. 423-326, The Hague, The Netherlands, August 30-September 3 (1992)

14. P.E. Danielsson: Euclidean distance mapping. Computer Graphics and Image Processing 14, 227-248 (1980)

15. H. Blum: A transformation for extracting new descriptors of shape. Symp. on Models for the Perception of Speech and Visual Form, Boston, November 1964. Mit Press, pp. 362-380 (1964)

16. U. Montanari: A method for obtaining skeletons using a quasi-euclidean distance. Journal of ACM 15, 600-624 (1968)

17. M.P. Martinez-Perez, J. Jimenez, J.L. Navalon: A thinning algorithm based on contours. Computer Vision Graphics and Image Processing 39, 186-201 (1987)

18. A. Rosenfeld: Digital topology. Amer. Math. Monthly, pp. 621-630 (1979)

19. H. Tamura: A comparison of line thinning algorithms from digital geometry viewpoint", 4th Int. Conf. on Pattern Recognition, IEEE Comp. Soc. Press, Kyoto, pp. 715-719 (1978)

20. H. Yamada: Complete euclidean distance transformation by parallel operation. 7th Int. Conf. on Pattern Recognition, IEEE Comp. Soc. Press, Montréal, pp. 69-71 (1984)

21. I. Ragnemalm: Contour processing distance transforms. 5th Int. Conf. on Image Analysis and Processing, Positano (Italy), September 89, pp. 204-212 (1990)

22. L.Vincent L: Exact euclidean distance function by chain propagations. Conf. on Computer Vision and Pattern Recognition'91, June 3-6, Maui (Hawaii), pp. 520-525.

23. G. Borgefors: Distance transformations in digital images. Computer Vision Graphics and Image Processing 34, 344-371 (1986)

24. E. Thiel, A. Montanvert: Chamfer masks: discrete distance functions, geometrical properties and optimization. 11th Int. Conf. on Pattern Recognition, The Hague, The Netherlands, Aug.30-Sept.3, pp. 244-247 (1992)

25. E. Thiel, A. Montanvert: Improvment of chamfer distances for images analysis. Technique et Science Informatique 11, 9-41 (1992)

26. J.L. Pfaltz, A. Rosenfeld: Computer representation of planar regions by their skeletons", Comm. of ACM 10,119-125 (1967)

27. A. Rosenfeld, A.C. Kak: Digital Image Processing, Academic Press (1982)

28. C. Arcelli, G. Sanniti di Baja: Finding local maxima in a pseudo-euclidean distance transform. Computer Vision Graphics and Image Processing 43, 361-367 (1988)

29. C. Arcelli, G. Sanniti di Baja: A one-pass two-operation process to detect the skeletal pixels on the 4-distance transform. IEEE trans. on Pattern Analysis and Machine Intelligence 11(4), 411-413 (1989)

30. C. Arcelli, G. Sanniti di Baja: A width-independent fast thinning algorithm. IEEE trans. on Pattern Analysis and Machine Intelligence 7(4), 463-474 (1985)

31. E. Thiel, A. Montanvert: Shape splitting from medial lines using the 3-4 chamfer distance. Int. Workshop on Visual Form, May 1991, Capri (Italy), Plenum Press, New-York, pp. 537-546 (1992)

32. C. Arcelli, G. Sanniti di Baja: Shape splitting using maximal neighborhoods. 6th Int. Conf. on Pattern Recognition, IEEE Comp. Soc. Press, Munich, October 1982, pp. 1106-1108 (1982)

33. A. Montanvert: Medial line: graph representation and shape description. 8th Int. Conf. on Pattern Recognition, IEEE Comp. Soc. Press, Paris, October 1986, pp. 430-432 (1986)

34. J-M. Chassery, A. Montanvert: Géométrie discrète en analyse d'images. Editions Hermès (1991)

35. A. Montanvert, J-M. Chassery, C. Charles: Méthodes de décomposition d'images binaires appliquées à la trajectographie sous-marine. Traitement du Signal, Numéro Spécial Vision par Ordinateur 4(3), 195-203 (1987)

36. S.N. Shihari: Representation of three-dimensional digital images. Computing Surveys 13 (4), 399-424 (1981)

37. D. Meagher: Geometric modeling using octree coding. Computer Graphics and Image Processing 19(2), 129-147 (1982)

38. E. Bertin, J-M. Chassery: 3D Voronoi diagram: application to segmentation, 11th Int. Conf. on Pattern Recognition, The Hague, The Netherlands, August 30-September 3, pp 197-200 (1992)

39. E. Bertin, R. Marcelpoil, J-M. Chassery: Morphological algorihms based on Voronoi and Delaunay graphs: microscopic and medical applications. Image algebra and morphological, Image processing III, SPIE ed., San Diego, July 18-23, pp. 356-367 (1992)

40. F. Rolland, J-M. Chassery, A. Montanvert: 3D medial surfaces and 3D skeletons. Proceedings of the First International Workshop on Visual Form, May 1991, Capri (Italy), Plenum Press, pp. 443-450 (1992)

41. G. Borgefors: Distance transformations in arbitrary dimensions. Computer Vision Graphics and Image Processing 27, 321-345 (1984)

Perceptual Grouping for Scene Interpretation in an Active Vision System *

Bruno Zoppis and Gaëlle Calvary and James L. Crowley

Lifia Institut IMAG, 46, avenue Félix Viallet, 38031 Grenoble Cedex, France

Abstract. This paper decribes a technique for scene interpretation within a continuously operating active vision system, based on the use of a vocabulary of perceptual grouping procedures for detection and labeling of objects. We propose a taxonomy of complex features composed of edge lines, describe the algorithms to construct each type of grouping from simple edge lines, and analyse the complexity of this computation. To reduce the size of input data for complex groupings we introduce chaining of grouping operations.

1 Introduction

We are investigating perceptual grouping for an active vision system, the VAP system [Vap, 1990], the overall objective of which is to construct, and to dynamically maintain, a description of the scene as perceived in a continuous image sequence. In its current version, the description is based on 2-D edge-segments at different resolutions that are tracked to maintain temporal continuity [Discours, 1990].

Using edge segments for object recognition is seriously complicated by two well-known problems :

1. The edge-segments in the 2-D description that compose instances of known objects are mixed with a large number of edge-segments from other sources causing a combinatorial explosion for the recognition process.
2. The edge-segments are corrupted by various phenomena. The existence of segments is uncertain and the parameters of segments are subject to be affected by random noise.

The first problem potentially causes a combinatorial explosion for the recognition process. This has led us to use a "hypothesis and verification" technique. In this technique, a recognition procedure is invoked based on the current context. This recognition procedure generates hypotheses about objects in the scene and about their composition.

Both of these problems also can be substantially reduced by the use of more complex features as basis for recognition : complex features contain more information and usually are more specific than a single edge line.

---
* **This work was sponsored by the CEC ESPRIT Basic Research office as part of Project 3038 "Vision as Process.**

Following [Lowe and Binford, 1983, Lowe, 1985, Veillon et al., 1990], we propose grouping of edge lines as such complex features. Our strategy is to provide a large vocabulary of perceptual grouping procedures that may be used to interrogate the 2-D image description. These procedures are called on the basis of expectations.

This paper proposes a taxonomy of complex features composed of edge lines, describes the algorithms to construct each type of grouping from simple edge lines, and analyses the complexity of this computation. Computational complexity, of course, is a crucial point for any component of a continuously operating vision system, since the results have to be delivered in a fixed cycle time. In order to satisfy the real time constraint, we must :

1. know the complexity of all used procedures,
2. eventually limit the amount of data treated in any one cycle,

and, first of all, design simple procedures, and implement them efficiently.

Grouping procedures are part of a larger vocabulary of image description access functions. These functions are organised into four classes, according to the nature and complexity of the operations that they perform. Grouping operations from the first two classes have complexities that depends on the number of primitives in the model, N, while grouping primitives from the third and fourth class have a quadratic complexity $O(N^2)$. These procedures may be composed in such a way that an $O(N)$ grouping procedure can be used to select the subset of model primitives that are considered by the more expensive primitives.

The second problem also requires techniques for dealing with uncertainty. Object hypotheses are accompanied by a confidence factor. Interrogation of the 2-D description yields a confidence factor, based on the similarity of the predicted and observed primitives. By combining evidence, missing components degrade the confidence of a label, but do not prevent its assignment. The hypothesis and verification process permits us to evaluate conflicting hypotheses in parallel.

2 A Taxonomy of Interrogation and Grouping Procedures

Model interrogation procedures may be organized into a taxonomy according to the nature of the interrogation that they make. We have designed four classes of interrogation procedures :

Class 1 Procedures for extracting parametric primitives that are close to an ideal geometric entity such as line or point. Such primitives are easily implemented and have a complexity that is proportional to the number of primitives in the model.

Class 2 Procedures for extracting primitives that are similar to an ideal "prototype". The prototype is specified as en estimated value and a standard deviation for each of the primitive parameters. Similarity is measured as a Mahalanobis distance. Parameters may be labeled as a "wild-card" by specifying a large (or infinite) standard deviation.

Class 3 Procedures for extracting pairs of primitives that satisfy some geometric relation, such as junction, an alignment, or a parallelism. The geometric relation is defined as a set of parameters and specified as a estimation and a standard deviation. These procedures have theoretical complexity of $O(N^2)$.

Class 4 Procedures for extracting grouping of groupings. If performed by brute force, complex forms such as lines and contours composed of sets of line segments can require an exponential number of computations. By structuring the interogation as a grouping, we can limit the theoretical complexity to $O(N^2)$.

In addition to these four classes it also is possible to interrogate the model using a composition of groupings. That is we can apply the grouping operations in sequence, for example, extracting all junctions of a certain form that are near an ideal line. Thus the $O(N^2)$ primitive can be used to restrict the number of entities interrogated (the value of N) by an $O(N^2)$ procedure.

The result of the interrogation is a list of segments or groups of segments that satisfy the specified geometric properties. These segments or groups are accompanied by a "quality factor" Q based on the similarity between the requested and observed entity, the list of ID's for the entities that satisfy relation is returned to the calling module, and a small set of properties of the grouping. This list is sorted based on the quality factor.

For each procedure, we will present its computational cost in the following terms :

A Number of additions or substraction per operation.
M Number of multiplications or divisions per operation.
C Number of comparisons per operation.
N Number of segments in the model.

All computational costs are worst case. The worst case often being quite rare, actual code contains a number of optimization to reduce the cose for average cases.

3 Parametric Representation for Line Segments

Edge-segments are computed by chaining gradient edge-points in a one pass algorithm [Chehikian, 1991]. Resulting segments are oriented according to the direction of gradient as shown in Fig.1.a.

The maintenance of the dynamical model can be simplified by using a representation composed of parameters for the midpoint, half-length and spatial extent [Crowley et al., 1988]. Such a representation also is useful for many of the matching operations involved in grouping and model access.

Tracked edge-segments in our 2-D description are expressed by a set of four independent primary parameters, and a set of dependent parameters, as shown in Fig.1. The primary parameters for an edge-segment are given by the vector $S = P_m, \theta, h$, where :

P_m is the midpoint expressed as (x, y).

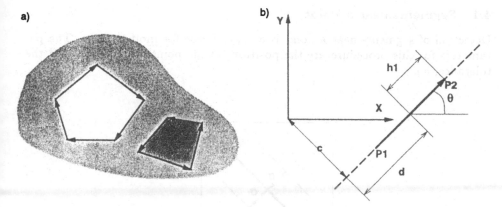

a) *Orienting segments according to gradient direction.*
b) *Midpoint-Distance-Length representation for segments.*

Fig. 1. Segment Representation

θ is the orientation of the segment ($[0, 2\pi[$).
h is the half length of the segment.

Each of these parameters is accompanied by a temporal derivative and a co-variance maintened by tracking [Crowley et al., 1988]. A number of dependent parameters that are used in many of the grouping operations are computed when needed from the primary parameters and then saved within the edge segments. These parameters are :

P_1, P_2 The end-points.
A, B Line Equation Coefficient ($A = sin(\theta), B = -cos(\theta)$)
C Perpendicular distance to the origin.
D Distance from projection of origin to midpoint.

Segments also include a confidence factor, CF, and a unique identity, ID.

4   Class 1 - Grouping with Respect to a Point or Line

This is the first class of procedures for extracting parametric primitives that are close to an ideal geometric entity such as a line or a point. Such primitives are easily implemented and have a complexity that is proportional to the number of primitives in the model. At the current time, procedures exist to retrieve a list of segments near a point in the image, or near to a line equation.

4.1 Segments near a Point

Detection of segments near a point is a typical case for model access. The parameters for this procedure are the positions of the point, $P = (x, y)$, and the tolerance, σ_d.

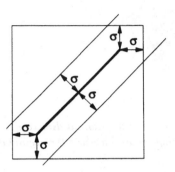

Fig. 2. Tests to detect segments near a point.

A segment is close to a point if the point is within a bounding box defined by the segment, and if the point has a distance of less than σ_d to the segment. The bounding box is defined by adding and substracting the tolerance σ_d to the endpoints of the segment. This operation requires 4 additions and 2 comparisons. The point is then tested for inclusion within this rectangle, requiring at most 4 comparisons. If the point is in the box, its perpendicular distance to the segment is computed using : $d_p = |Ax + By + C|$

In the worst case, where the point is within the bounding box of every segment, the computational cost for N segments is :

$$Cost = (4A + 2M + 6C)N$$

The edge segments that pass the test are assembled into a list sorted according to a quality factor. The quality factor is given by the formula :

$$Q = \frac{d_p}{\sigma_d}$$

A segment with a quality factor of zero is ideal. The sorted list of segments ID's with their quality factor is returned.

4.2 Segments near a Line

The parameters for extracting segments near a line are the coefficients of the line equation, (A, B, C), and the tolerance distance, σ_d. Three values are compared to the tolerance distance : The perpendicular distance to each of the end-points, and the intersection point between the line and the segment. If the segment intersects the line, then its quality factor is zero (fig 3).

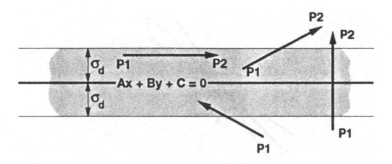

Fig. 3. Four cases of segments near a line.

The distance from the end-points d_1 and d_2 of a segment to a line may be computed from the line coefficients, A, B, C by the formula : $d_i = AX_i + BY_i + C$. Each distance requires 2 multiplies and 2 adds. The overlap of the segment with the line equation requires comparing the sign of d_1 and d_2. If their signs are different, then the segment traverses the line, and the distance to the line is taken as zero. Thus the cost of evaluating this function for each of N model segments is :

$$Cost = (2M + 2A + 3C)N$$

The quality factor is based on distance from the line compared to the tolerance σ_d :

$$Q = \frac{min(d_i)}{\sigma_d}$$

The ID's of the segments that pass the test are inserted in a list with a position based on their quality factor. The sorted list of segment ID's with their quality factor is returned.

5 Class 2 - Grouping with Respect to a Prototype

The second class of model interrogation procedures are for edges that are similar to an ideal "prototype". The prototype is specified by an estimated value and a standard deviation for each of the primitive parameters. similarity is measured by a Mahalanobis Distance. Parameters may be labeled as a "wild-card" by specifying a large (or infinite) standard deviation. This test is illustrated in the Fig.4.

The prototype segment is given by parameters $(x_p, y_p, h_p, \theta_p)$. A tolerance vector for the parameters is given by $(\sigma_x, \sigma_y, \sigma_\theta, \sigma_h)$. The Mahalanobis distance for each parameter is given by square of the difference divided by the square of the tolerance. We can perform an equivalent test by simply comparing the absolute value of the difference to the tolerance (where abs() has a cost of one comparing). In the worst case, for N segments, this test then costs :

$$Cost = (4A + 8C)N$$

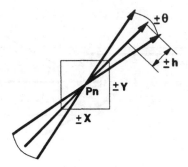

Fig. 4. Similarity with respect to a prototype segment.

For each segment that passes all four tests, the quality is given by :

$$Q = \left(\frac{x_p - x}{\sigma_x}\right)^2 + \left(\frac{y_p - y}{\sigma_y}\right)^2 + \left(\frac{h_p - h}{\sigma_h}\right)^2 + \left(\frac{\theta_p - \theta}{\sigma_\theta}\right)^2$$

The ID's of the segments that satisfy all four tests are inserted in a list based on their quality factor. The sorted list of segment ID's with their quality factors is returned.

6 Class 3 - Grouping of Pairs of Segments

The simplest groupings of segments are pair-wise groupings. Procedures for extracting pairs of primitives that satisfy some geometric relation, such as junction, alignment, or parallelism, constitute the third class of model access procedures. The geometric relation is defined as a set of parameters and specified as an estimation and a standard deviation. Detection of pair-wise groupings has an inherent complexity of $O(N^2)$.

6.1 Junctions of Two Segments

A widely used pair-wise grouping are junctions of two segments. Junctions are defined with three parameters, θ, ψ and d, as shown in Fig.5.
    A prototype junction is specified as :

ψ, σ_ψ : The internal angle of the junction and its tolerance.
θ, σ_θ : The orientation of the junction and its tolerance.
σ_d : The tolerance of the distance between end-points.

For each segment, the procedure tests every segment (other than itself) and computes the square of the distance from the point P_2 of the first segment to the point P_1 of the other segment. This operation costs $3A + 2M$. If the squared

190

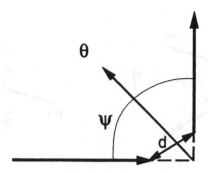

Fig. 5. A Pair-wise Junction of two segments.

distance is less than the square of the tolerance, then the internal angle and external angle for segments 1 and 2 are computed as :

$$\psi_s = \theta_1 - \theta_2$$

$$\theta_s = \theta_1 + \theta_2$$

6.2 T-Junctions

A T-junction is another common class of pair-wise junction. A T-junction has the same parameters as a pair-wise junction. A prototype T-junction is specified as :

ψ, σ_ψ : The internal angle of the junction and its tolerance.
θ, σ_θ : The orientation of the junction and its tolerance.
σ_d : The tolerance of the distance between end-points.

As with junctions, the internal angle and external angle are computed as :

$$\psi_s = \theta_1 - \theta_2$$

$$\theta_s = \theta_1 + \theta_2$$

These are two types of T-junction. Those formed with point P1 and those formed with point P2, as shown in Fig.6. For each segment, a list of candidate T-junctions is computed by testing the end-points of all other segments to see if they fall within the bounding box defined by adding σ_d to the segment end-points. For any end-point found in the box, the perpendicular distance to the support line of the edge segment is computed, and compared to the tolerance distance. Because this must be computed between all pairs of segments, the cost is :

$$Cost = (4A + 2M + 6C)N(N - 1)$$

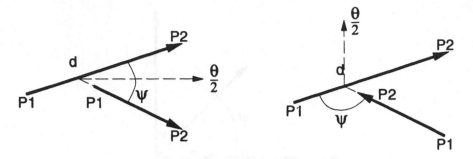

Fig. 6. Two types of T-junction and their parameters.

If the distance is less than the tolerance, the edge is kept in the list of T-junctions. The quality is given by :

$$Q = \left(\frac{d}{\sigma_d}\right)^2 + \left(\frac{\theta_s - \theta_k - \theta}{\sigma_\theta}\right)^2 + \left(\frac{\theta_s + \theta_k - \psi}{\sigma_\psi}\right)^2$$

Where d is the perpendicular distance, and θ_s is the orientation of the segment "s" that defined the box, and θ_k is the orientation of the segment whose end-point was in the bounding box. If one of the segment "s" is closer than σ_d to the end point being tested, then the pair of segments forms a normal junction, and not a T-junction.

6.3 Aligned Segments

Aligned, parallel and anti-parallel are a mutually exclusive relations between two segments as indicated in Fig.7. These relations are based on three measurements :

1. The similarity of the orientation σ_θ,
2. The perpendicular distance of the midpoints, σ_c,
3. The distance between the projection of end-points to the supporting line, σ_d.

The parameters that define alignments are C,D and θ. Two segments S_1 and S_2 are said to be aligned if :

1. The two segments have similar orientation $((\widehat{S_1, S_2}) < \sigma_\theta)$,
2. Each segment is near the support line of the other one $(|C_1 - C_2| < \sigma_c)$, and
3. The end-point P_1 of one segment projects to the support line of the second segment a distance less than σ_d beyond the end-point P_2 of the second segment.

For each segment S, the tests are applied to every other segment k. The first test is easily made by comparing the parameters θ. The difference in orientation

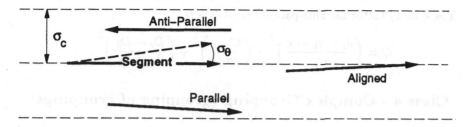

Fig. 7. Parallel, Anti-parallel, and aligned segments.

must be less than σ_θ . The second test is based on the perpendicular distance
from the end-points of segment k to the support line of segment $S : C_1$ and C_2.
If C_1 and C_2 have opposite signs, the segment crosses the support line, and the
distance is $C_{min} = 0$. Otherwise, the test is based on $min(|C_1|, |C_2|)$.

If $C_{min} < \sigma_c$ the test is passed. This test requires $4M + 4A + 3C$ for each
segment.

The final test is based on where the end-points of the second segment project
onto the support line. This projection can be expressed in terms of a parameter
D, that is the distance from the projection of the origin onto the support line.
For a line equation, $Ax + By + C = 0$, the D parameter is computed by :
$D = Bx + Ay$

As each segment s is considered, we compute D_s for the midpoint of the
segment s. To determine the overlap of a segment k with the segment s, we use
the parameters (A,B) of s to compute the parameter D_k for the midpoint of the
segment k. Three case are possible based on the half-length, h, of the segments
s and k :

Case 1 : $D_s + h_s < D_k - h_k$ Segment k is aligned after segment s.
Case 2 : $D_s + h_s > D_k - h_k$ Segment k is aligned before segment s.
Case 3 : Otherwise   Segment k and segment s overlap.

For the procedure to detect alignment, only case 1 of test 3 is applied. This
give a cost for the third test of $(2M+A)N^2$. The total cost of detecting alignment
is :

$$Cost = (6M + 5A + 4C)N^2$$

In case 1, the segments are marked as aligned. In case 2, the relationship
is not marked. It will be detected when segment k is compared to all other
segments.

Anti-parallel segments are similar to parallel segments, except that the difference in orientation should be close to π.

The quality factor for aligned segments is :

$$Q = \left(\frac{\theta_s - \theta_k}{\sigma_\theta}\right)^2 + \left(\frac{C_{min}}{\sigma_C}\right)^2 + \left(\frac{D_s - D_k + h_s + h_k}{\sigma_D}\right)^2$$

The quality factor for parallel segments is :

$$Q = \left(\frac{\theta_s - \theta_k}{\sigma_\theta}\right)^2 + \left(\frac{C_{min}}{\sigma_C}\right)^2 + \left(\frac{D_s - D_k}{h_s + h_k}\right)^2$$

The quality factor for anti-parallel segments is :

$$Q = \left(\frac{\theta_s - \theta_k - \pi}{\sigma_\theta}\right)^2 + \left(\frac{C_{min}}{\sigma_C}\right)^2 + \left(\frac{D_s - D_k}{h_s + h_k}\right)^2$$

7 Class 4 - Complex Grouping (grouping of groupings)

Detecting lines and contours in the presence of multiple gaps is notoriously hard. If attacked in a naive way, this process has an exponential complexity. By performing the detection as a group of groups, such forms may be detected with a theoretical maximum complexity which is in $O(N^2)$ and which is generally much smaller.

7.1 Y-Junctions

Y-junction are built up as pairs of pair-wise junctions. All pair wise junctions involve the proximity of a head (P_2) with a tail (P_1) of a second segment. Only two types of Y-junctions are possible : Y-junctions with two tails (fig 8.a) or with two heads (fig 8.b).

The parameters for the Y-junction are :

Type : two heads or two tails.
ψ_1, σ_{ψ_1} : internal angle of the first junction.
ψ_2, σ_{ψ_2} : internal angle of the second junction.
θ, σ_θ : sum of the orientations of the junctions.
σ_d : tolerance of the distance between points of the two junctions.

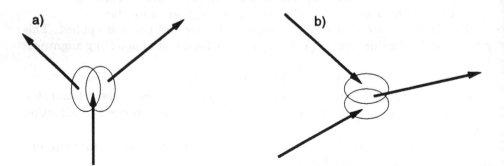

a)

b)

Fig. 8. The two types of Y-Junction.

A Y-junction is detected based on the coincidence of a segment in two junctions. That is, all two segment junctions are detected using the specified parameters for the first junction. All specified parameters for the second junction are then detected. Each operation returns a sorted list of junctions that includes the ID's of the segments. For each junction in the first list, the second list is scanned to search for the any junction that includes one of the two segments. For each case where the two junctions contain a common segment, each parameter is compared to its tolerance. If each parameter is within the tolerance, then the quality factor is computed as :

$$Q = \left(\frac{d_1}{\sigma_{d_1}}\right)^2 + \left(\frac{d_2}{\sigma_{d_2}}\right)^2 + \left(\frac{\psi_1 - \psi_{s_1}}{\sigma_{\psi_1}}\right)^2 + \left(\frac{\psi_2 - \psi_{s_2}}{\sigma_{\psi_2}}\right)^2 + \left(\frac{\theta_p - \theta_s}{\sigma_\theta}\right)^2$$

The cost of the operation is :

$$Cost = 2(7A + 2M + 3C)N(N - 1) + CN^2$$

7.2 Complex Edge Lines

A common problem in edge segment description is that segments break in unpredictable places. Sequences of broken segments that are aligned can be grouped to form longer segments using a complex grouping operation. The parameters for the procedure for detecting a complex alignment are the same as those for detecting a simple alignment.

σ_θ : tolerance for deviations of orientations within a segment.

σ_c : tolerance for perpendicular displacement.

σ_d : maximum gap between points.

The resulting composite line segment has the same attribute as a simple line segment.

The algorithm begins by constructing a list of all pair-wise alignments of segments using the parameters σ_θ, σ_c, σ_d and the algorithm for detecting alignments described above. As before the cost of this step is $(6M + 5A + 4C)N^2$. The result is a list of pairs that are aligned, as shown in Fig.9 .

Each pair is assigned a "pair-ID", to give a list L composed of triples : $(Pair - ID, (Seg - ID_1, Seg - ID_2))$. This list of segment pairs then is scanned to detect alignment pairs that share a segment. This test is based uniquely on matching the Seg-ID's.

To test the alignment of two groups, a composite edge segment is computed by taking the end-point P_1 of $Seg - ID_1$ and the point P_2 of $Seg - Id_2$. The midpoint and orientation for this composition are computed. We then make exactly the same tests as for alignment of two segments. If the composite segments are aligned, with respect to the specified tolerances, the two groups are retracted from L and a new group is appended to the end of L. The procedure is repeated until no further groupings can be formed.

7.3 Arcs

A procedure similar to that for complex edges can be used to detect complex arcs. This procedure is based on grouping junctions of two segments. The parameters for detecting an arc are :

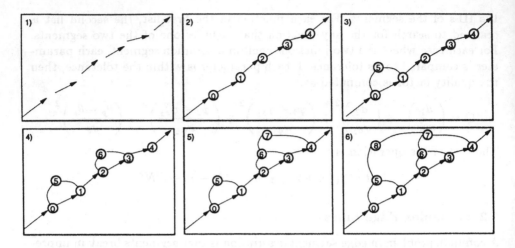

1) *Set of initial segments.*
2) *Constructed list of pair-wise alignments of segments.*
3) *Beginning of the research, a new group is formed with groups 0 and 1 that share one segment.*
4,5) *Continuation of the research.*
6) *Final grouping.*

Fig. 9. Construction of a complex grouping

σ_d : Maximum gap between end-points.
K : Desired curvature.
σ_K : Tolerance for deviation in curvature.

The procedure returns :

P_c : The center point of the composite arc.
K : The average curvature of the arc.
P_1, P_2 : The end-points of the arc.

Our algorithm for detecting edge segments approximates an arc with a segment whose end-points tend to lie on (or close to) the arc. The largest errors tend to be at the center of the segment. Thus an arc is defined as passing through the end-points of segments.

The bisector of a segment defines the locus of centers of circles passing through the end-points of a segment. If two segments share a junction, and are not parallel, their bissectors will have an intersection point. The intersection point defines a common center-point for the arcs for each of the segments, as illustrated in Fig.10. The radius of each arc can be computed as the distance of

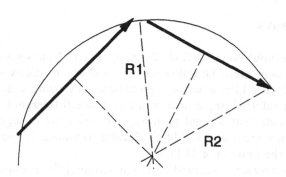

Fig. 10. An arc grouping.

either end-point of the segment to this center-point. The two radii will not be necessarily equal.

To arrive at a common radius, we can calculated a weighted average of the two radii, using the segments length, s, defined by the radius and the angle subtended by the arc, α. Thus, for segments 1 and 2, the joint radius is given by :

$$R = \frac{1}{S_1 + S_2}(R_1 S_1 + R_2 S_2)$$

Two arcs form a complex arc if they share a segment. To decide if they are part of the same arc, we compare the difference in curvature to the tolerance σ_K.

The quality factor is based on the difference between the desired and observed curvature :

$$Q = \frac{K_s - K}{\sigma_K}$$

8 Composition of Grouping Operations

The first two classes of operations described above each produce a list of ID's of edge segments that satisfy some relation. The third class produces a list of pairs of edge-segments. The fourth class produces a list of arcs, curves or Y-Junctions, each of which contains a list of segments. Each of these procedures contains a "mode" argument that will cause them to return the list of edge segments that were detected. The list of segments produced in this mode can be substituted for the edge segment model in subsequent model access procedures. This makes it possible to compose a chain of model access operations, using the list of segments from each operation as input to the next.

Chaining model access operations is a form of logical "AND" operation. For example, with such a chain, it is possible to request all of the junctions of two segments that are near an ideal line segment, or all anti-parallel lines near a point. Chaining operations permit us to use $O(N)$ operations from classes 1 and 2 to select the candidate segments considered by the more expensive $O(N^2)$ operations of classes 3 and 4.

9 Experiments

Fig.11.a shows an input image (512x512 pixels). Fig.11.b shows all the segments extracted from this image. The following figures show some examples of grouping procedures. In Fig.11.c, squares are extracted by using pair-wise junction grouping with an internal angle near $\pi/2$ and a small tolerance on this angle. In Fig.11.d, segments near a point grouping produce a set of segments that are used as input for arc grouping (Fig 11.e). Similar treatment is applied to extract arc groupings on the plate (Fig 11.f).

Table 1 shows average measured time (in seconds) for grouping procedures on a Sun3 workstation. We can observe that to improve performance we must use restrictive tolerances or reduce the number of input segments. "Hypothesis and verification" permits us a good parameter estimation for groupings and thus to use small tolerances. Chaining operations permit to reduce the set of segments which are the input for complex groupings.

Table 1. Average measured time for grouping procedures.

Number of segments	100	150	200	250	300	350
Near a point	0.02/0.025	0.02/0.035	0.02/0.05	0.02/0.06	0.02/0.08	0.02/0.1
Near a line	0.02/0.02	0.02/0.02	0.02/0.02	0.02/0.02	0.02/0.02	0.02/0.02
Prototype	0.02/0.045	0.02/0.05	0.02/0.06	0.02/0.06	0.02/0.075	0.02/0.1
Aligned	0.02/0.045	0.04/0.06	0.06/0.1	0.1/0.16	0.12/0.22	0.16/0.3
Junction	0.02/0.045	0.02/0.1	0.025/0.14	0.04/0.16	0.045/0.2	0.06/0.22
T Junction	0.02/0.1	0.05/0.25	0.1/0.4	0.25/0.55	0.2/0.8	0.25/0.9
Y Junction	0.04/0.2	0.04/0.5	0.05/0.65	0.06/0.7	0.07/0.8	0.1/0.9
Arc	0.03/0.1	0.04/0.14	0.04/0.2	0.06/0.3	0.07/0.6	0.08/0.75

This table presents average measured time in seconds for grouping procedures according to the number of input segments. The left number is the time for restrictive tolerances in the groupings, and the right number is the time for infinite tolerances. The value 0.02 is the smallest value given by the timer and not a significant execution time.

10 Conclusion

Studies in "gestalt psychology" [Bruce and Green, 1990] have shown the significance of visual primitive groupings for scene interpretation. We have described an approach of these groupings as an associative access to dynamical model of the primitives. We also have proposed a taxonomy of these groupings.

Some previous efforts at perceptual grouping had based the computations on the use of a pixel labeling to represent the graph of the edge segments [Veillon et al., 1990]. Our experience is that such representations are inappropriate for continuous operation. In any edge based description of an image there always are gaps along the edge segments and at junctions. Closing such gaps

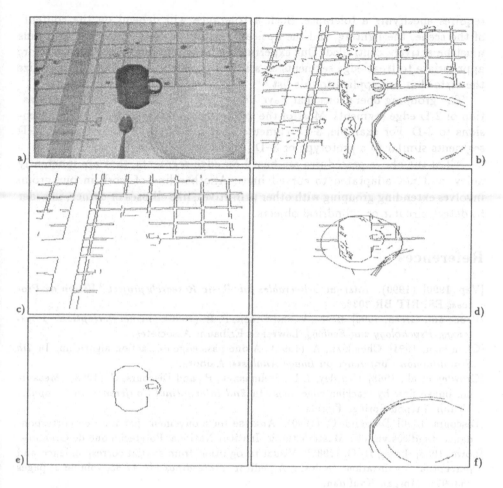

a) original image.
b) extracted segments.
c) Pair-wise junctions of two segments with an angle near $\pi/2$
d) Segments near a point. The ellipse represent research area.
e) Arc groupings on the segments extracted in d.
f) Arc groupings on the segments near plate center.

Fig. 11. Use of grouping operations on an image.

requires specifying a tolerance. Such a tolerance will depend on the contents of the scene, the lighting, and the optical parameters of the cameras, and thus must be actively controlled. Our experiments have convinced us that it is more appropriate to treat edge segments as a list of property vectors and to optimize the tests that are applied to this list.

The grouping operations that have been presented all concern the organization of 2-D edge segments. Most of the grouping operations have natural extensions to 3-D. For example, 3–D segments near a 3–D point, line or plane, 3–D segments similar to a prototype or 3–D junctions.

All of the above procedures are based on edge segments, that are inherently noisy, and not adaptated to curved lines. Another line of work in our group involves extending grouping with other primitives, like ellipses or corner, in order to detect circular or cylindrical objects.

References

[Vap, 1990] (1990). *Internal Deliverables for Basic Research project "Vision as Process*. ESPRIT BR 3038.

[Bruce and Green, 1990] Bruce, V. and Green, P. R. (1990). *Visual Perception – Physiology, Psychology and Ecology*. Lawrence Erlbaum Associates.

[Chehikian, 1991] Chehikian, A. (1991). A one pass edge extraction algorithm. In *7th Scandinavian Conference on Image Analysis*, Aalborg.

[Crowley et al., 1988] Crowley, J. L., Stelmaszyk, P., and Discours, C. (1988). Measuring image flow by tracking edge lines. In *2nd Internatinal Conference on Computer Vision*. Tarpon Springs Florida.

[Discours, 1990] Discours, C. (1990). Analyse du mouvement par mise en correspondance d'indices visuels. Master's thesis, Institut National Polytechnique de Grenoble.

[Lowe, 1985] Lowe, D. G. (1985). Visual recognition from spatial correspondence and perceptual organization. In Joshi, A., editor, *proceedings IJCAI 85*, volume 2, pages 953–959. Morgan Kaufman.

[Lowe and Binford, 1983] Lowe, D. G. and Binford, T. O. (1983). Perceptual organization as a basis for visual recognition. In *proceedings AAAI-83*, pages 255–260. William Kaufman, Inc.

[Veillon et al., 1990] Veillon, F., Horaud, P., and Skordas, T. (1990). Finding geometric and relational structures in an image. In *1st European Conference on Computer Vision*, Antibes. Springer Verlag.

Incremental Free-Space Modelling From Uncertain Data by An Autonomous Mobile Robot

Philippe Moutarlier and Raja Chatila

LAAS - CNRS
7, Avenue du Colonel Roche
31077 Toulouse Cedex, France

Abstract. Explicit representation of free-space is necessary for mobile robot navigation. This paper presents a fully implemented system that achieves incremental environment modelling for a 2D mobile robot using a laser range-finder. This system produces a surfacic model of objects (polylines and associated uncertainties) and a model of free-space (polygons) taking into account possible appearance and disappearance of objects.

1 Introduction

In order to plan collision-free paths, a mobile robot needs a representation of free-space. We consider in this paper a mobile robot that explores its environment and perceives it with a non-perfect sensor, and has to build its free-space representation *incrementally*. Such a robot has to autonomously:

- represent object surfaces using uncertain data from local perceptions;
- update and correct its position with respect to environmental features;
- fuse local object models as well as free-space representations to build global models;
- select adequate viewpoints that take into account sensor limitations;
- plan and execute collision-free trajectories in order to reach these viewpoints.

We have developed, implemented and experimented on the Hilare mobile robot such a complete system. This paper focuses on the problem of free-space representation.

Several authors have addressed the problem of building surfacic representations from uncertain sensor data [1, 2, 3, 4, 14], and we have developed in this context [12] an approach for incremental environment 2D map making using a laser-range finder. However, in a partially known environment, the representation of object surfaces is not sufficient *by itself* for computing the free-space. Indeed in this case, objects have open contours and their connection to delimit free-space is not unique (figure 1). Hence the robot must also keep a representation of the natural connection provided by the surface (in 2D) or the volume (in 3D) swept by the sensor while acquiring the data on object surfaces.

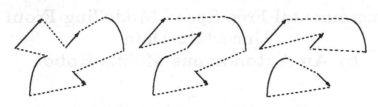

Fig. 1. Example of three different possible free-space representations from the same surfacic data.

In order to represent free-space, several authors use a grid representation of free and occupied space [13, 8, 7, 5] and do not use an algebraic representation of object surfaces and frontiers of free-space. In addition to occluding possible paths because of grid spacing, such approaches (including certainty grids [9]) do not provide for an explicit representation of uncertainties on the position of object surfaces as such. And it is important to notice that Object surfaces, considered as features, and their uncertainties are needed for robot localization. Iijima [6] makes use of two models: an analytical model for representing surfaces and updating them by fusion, and a grid model for representing space.

Our approach is to build an algebraic representation of free-space, consistent and compatible with the surfacic representation. This problem is addressed in the general case where:

- Measurements are imprecise. The surfacic representations have to be updated to their best estimates at each perception. This can cause inconsistencies in the free-space representation.
- Measurements are incomplete. Sensors have their limitations (e.g. specularities). A sensor model is necessary to predict situations where existing objects may be not perceived.
- The environment is not static. In the real world, objects may move, and we have to take into account their appearance and disappearance, and the consequences on free-space.

The following section presents an overview of the approach that was actually implemented and experimented on the Hilare mobile robot for the incremental representation of free-space. The sensor is an azimuth scanning laser range-finder, and the models are therefore 2D. Piecewise linear approximations are built from laser range data, and the object models are polylines. Section 3 briefly summarize the construction of surfacic representations. In sections 4 and 5, we present the construction of free space respectively in the special case of a perfect sensor and in the general case that takes into account sensor inaccuracies and limitations. Experimental examples are provided to illustrate the approach.

2 Uncertain Free-space Incrementation

Let us first decribe the general process of new data integration.

At instant k and before the acquisition of new data, the robot has the following models (fig 2): i) The Global Surfacic Model (GSM_k) that is the representation of the actual object surfaces currently known, ii) The Global Free-Space model (GFS_k) representing the current estimation of the free-space. A step of data integration at instant $k+1$ can be described as follows (arrow labels on figure 2 correspond to the four following operations):

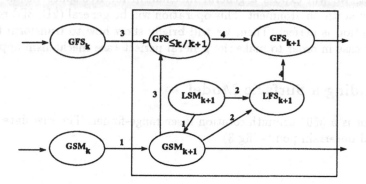

Fig. 2. A step of model incrementations after data acquisition

1. Sensors (laser range-finder, ultrasonic sensors, stereovision) generally give information on object surfaces which we call LSM_{k+1}, the Local Surfacic Model at instant $k + 1$. Our local model is a set of polylines with their extremities and the variance on their supporting line parameters (see [11] for details). This probabilistic representation enables us to update both the robot's position estimation and the supporting line parameters of the surfaces that were known before. The system stamps the date $k+1$ on the newly detected objects. This process provides the model GSM_{k+1}. It will be the subject of section 3.

2. There is generally no obstacles between the sensor and the measured surfaces. It is possible to build, from the surfacic data, a Local Free-Space representation (LFS_{k+1}) of the space between the sensor and the perceived objects. This operation will have to take into account sensor limitations by reasoning on the perceptibility of the objects in the Global Surfacic Model.

3. After the first operation decribed above, the surfaces of GSM_{k+1} belonging to GFS_k do not have the same parameter values in the free-space and the surfacic models. In order to keep the consistency between our representations, we propagate the surfacic modifications to the free-space boundary.

This step produces $GFS_{\leq k/k+1}$ (this notation is explained in section 4. Operations (2) and (3) will be discussed in section 5

4. Now, local and global free-space are exactly superimposed with the global model of surfaces and it is possible to compute the resulting free-space and to deal with moving objects by reasoning on time stamps. The new Global Free-Space model is thus produced. This will be the subject of section 4. The robot is then ready to decide to move in order acquire new information about its environment.

Our main purpose in this paper is the free-space fusion aspect of this general process. However, its necessary to define what we call "Global Surfacic Model". After that, we will expose a general mechanism for free-space fusion dealing with a non-static environment. This operation will be general (2D, 3D) provided the perception is perfect. Then, we will briefly show how to transform the real imperfect case in order to make the general perfect case mechanism applicable.

3 Building a Surfacic Model

Our sensor is a 360° azimuth rotation laser range-finder. The raw data is a set of ordered uncertain points (fig 3).

Fig. 3. The robot and the laserrange-finder raw data

Knowing the variance on each point estimation, points may be grouped into line segments with their variances. The result of this local segmentation step is shown in figure 4.a. The resulting model is called LSM_{k+1} the Local Surfacic Model at $k+1$. The integration of new surfacic data can be summarized as follows:

1. The correspondances between LSM_{k+1} and GSM_k, the global surfacic model GSM_k up to instant k, are computed;
2. then, each couple of matched segments is used to:

(a) relocate the robot's position estimation (fig 4.b and 4.c)

(b) update the surfacic parameters of GSM_k by two operations: a) probabilistic estimation of the supporting lines using stochastic fusion process [11], and b) computing the maximum extension of matched segment superimposition (fig 4.d)

New objects (not matched) of LSM_{k+1} are added to the global surfacic model producing GSM_{k+1}. We present next the construction of free space.

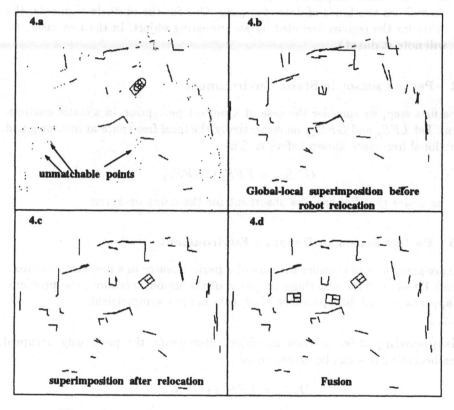

Fig. 4. A complete step of surfacic fusion

4 The perfect sensor case

4.1 Notations

We will use the following notations and definitions in the next sections:

1. $E_{<k}$ and $E_{\leq k}$ will respectively denote the free-space known up to instant k, before and after integration of the information acquired at k (i.e., new parts or disappeared objects).

2. An observation provides information about already known objects and possibly on previously unknown space. We call $E_{<k/k}$ the result of the updating of $E_{<k}$ by observation k.
3. Object surfaces are denoted S_o and boundaries with the unknown space (lignes of sight, for example) S_i.
4. $\tilde{E}$ is the boundary of E.
5. $C_X(X \cap Y)$ (the complementary set in X of $X \cap Y$ will be noted $X \ominus Y$.
6. when a space E is composed of several disconnected regions, we will note E^i a connected component of E.
7. we will use two kinds of date stamping. One for the obstacle segments, the other for the regions occluded by an appearing object. In the two cases, we will note it **date()**.

4.2 Perfect sensor in Static Environment

As a first step, we consider the case of a perfect perception in a static environment. Let LFS_p and $GFS_{<p}$ be respectively the local free-space at instant p and the global free-space known before it. Then:

$$GFS_{\leq p} = LFS_p \cup GFS_{<p}$$

Let us notice that free-space is absorbent for the union operation.

4.3 Perfect Sensor in Dynamic Environment

We are going now to consider the case of a perfect sensor in a non-static environment. Because of the absorbing property of the union operation, the problems of appearance and disappearance of objects are not symmetrical.

Disappearing objects When an object disappears, the previously occupied area becoming free can be expressed as :

$$D_{<p/p} = LFS_p \ominus GFS_{<p}$$

The union operation described above will add automatically this area to the free-space.

Appearing objects This case is more complex and will be described with the support of figure 5.

The appearance of a new object transforms a free-space region into an unknown region. The set of these newly unknown areas is included in (see right part of figure 5.b) :

$$A_{<p/p} = GFS_{<p} \cup LFS_p \tag{1}$$

However this last relationship represents also simple occlusions, and not only newly appearing objects. Date stamping of surfaces will allow us to filter the

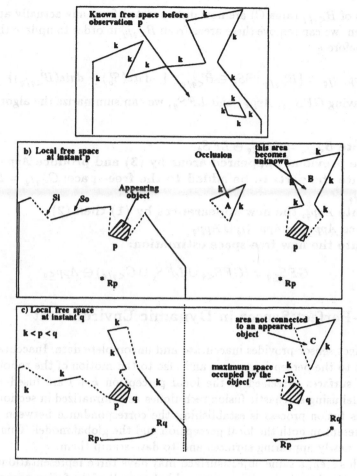

Fig. 5. Free-space updating in the case of moving objects
(observations at instants p and q).

actual appearance. Indeed, having $\{A^i_{<p/p}\}$ the connected component partition
of $A_{<p/p}$, the actual appearing objects at instant p can be detected by:

$$App_p = \{A^i_{<p/p} : \exists S^j_o \in \breve{A}^i_{<p/p} \mid date(S^j_o) = p\} \qquad (2)$$

Then, each connected component App^i_p of App_p is date stamped p.

After a robot movement around such a kind of object (area B in the figure), a new perception will improve its knowledge about it. Let $B_{<q/q}$ be the space representing the appearance before q from which the parts perceived at q are removed (areas C and D of the figure). Calling $App_{<q}$ the union of regions occluded by some appearing objects before q, we have

$$B_{<q/q} = App_{<q} \ominus LFS_q$$

Some parts of $B_{<q/q}$ (area C) are now disconnected from any actually appeared object. Then, we can remove these areas from $B_{<q/q}$ in order to update the areas appeared before q:

$$App_{<q/q} = \{B^i_{<q/q} : \exists S^j_o \in \breve{B}^i_{<q/q} \mid date(S^j_o) \geq date(B^i_{<q/q})\} \qquad (3)$$

Finally, having $GFS_{<q}$, $App_{<q}$ and LFS_q, we can summarize the algorithm as follows:

1. **compute** $B_{<q/q} = App_{<q} \ominus LFS_q$
2. **update previously appeared areas by (3) and produce** $App_{<q/q}$.
3. **compute the parts to be added to the free-space:** $C_{<q/q} = B_{<q/q} \ominus App_{<q/q}$
4. **compute** App_q, **the new appearances by (1) then (2)**
5. **produce** $App_{\leq q} = App_{<q/q} \cup App_q$
6. **compute the new free-space estimation:**

$$GFS_{\leq q} = (GFS_{<q} \cup LFS_q \cup C_{<q/q}) \ominus App_{\leq q} \qquad (4)$$

5   Non-Perfect Sensor in Dynamic Environment

A non-perfect sensor provides inaccurate and uncomplete data. Inaccuracies are due in fact to the perception sensor and also to the motion of the robot.

Object surfaces perceived at the local perception $k + 1$ are fused with the global model using stochastic fusion techniques as summarized in section 3. One step of this fusion process is establishing the correspondance between surfaces that are present in both the local perception and the global model. This permits to identify newly appearing surfaces and to date-stamp them.

After this step, a same object surface may have three representations: a) the surface perceived at the local perception, b) the result of the fusion process if this surface was already in the global map, and c) its representation as a boundary of free-space in the free-space model. We will present in the next section our method to compute a free-space representation having the same object surface parameters as in the surfacic model GMS_{k+1}.

5.1   Building the Local Free-Space

Building the local free-space relies upon a space interpolation between successive surface measurements (figure 6). This interpolation makes the assumption that there is an arrangement of the measurements such that there is no obstacle within the space between two successive measurements.

In the case of the 2D rotating laser range-finder, this arrangement is naturally given by the scanning order. The local surfacic model is then composed of three kinds of data ordered along the sensor rotation: a) successive aligned points grouped into line segment (lines S) with their uncertainties, b) sets of successive non-aligned points unmatchable together within a segment (e.g., area P) and

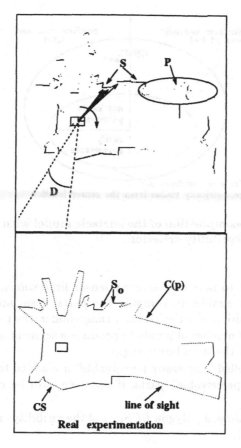

Fig. 6. Local perception and associated free-space

c) angular sectors (sector D) without measurements, that may correspond to specular surfaces.

Each kind has its interpretation as a boundary of the local free-space: a) a line segment is used directly as a free-space obstacle frontier (S_o); b) the frontier due to a set of points such as P is the part of the convex hull of P facing the sensor $(C(p))$. The corresponding polylines are labelled S_i c) where no measurements are available, we examin the corresponding sector in order to detect a possible previously known but non-perceived object (e.g belonging to GSM_k). If it is not the case, we choose to close the local free-space polygon with a closing segment (CS).

In order to have a unique surfacic representation, the computation of these sets is based on reasoning on geometric visibility and sensor perceptibility (fig 7).

The geometrical visibility area from a point within a polygonal space gives the boundary parts that may be visible from the sensor location. Futhermore, this operation allows to get directly the surfacic parameters from GSM_{k+1}.

209

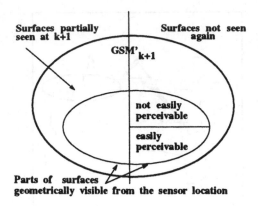

Fig. 7. Decomposition of the surfacic model with respect
to a perceptibility criterion

In addition, in order to take into account sensor limitations, a **perceptibility model** of the sensor enables to compute the parts of surfaces not perceivable from the sensor's position. For the 2D laser range-finder, the perceptibility model includes the minimal number of points to produce a segment approximation, the maximum range and the specularity angle.

The surfaces labelled "not easily perceivable" are added to the local surfacic model. The "easily perceivable" parts, if not seen, will be considered to have disappeared.

Finally, the star-shaped polygon LFS_{k+1} of the visibility area from the sensor location is computed.

5.2 Updating the previous Free-Space representation

In order to obtain an exact superimposition of the local and global free-space, we have now to update GFS_k. One problem is then to determine the endpoints of the boundaries in accordance with the surfacic model. Two solutions are possible: a) we can make an orthogonal projection of the global free-space extremities on the corresponding supporting lines in GSM_{k+1}, or b) we may also directly assign the extremity values of GSM_{k+1} to the corresponding segment in GFS_k.

The first solution always gives a shorter length than for the original segment and cannot converge to the actual length. The second method sometimes leads to inconsistencies (fig 8 represents an actual experimentation) in the intersections of boundary lines[1]. Provided we filter this kind of intersections, this method can converge to the actual segment length. We have therefore chosen this second alternative. We will not detail this complex aspect here[2].

This operation produces the model $GFS_{\leq k/k+1}$.

---

[1] notice that this case may also occur simply for numerical problems
[2] Full details are given in [10]

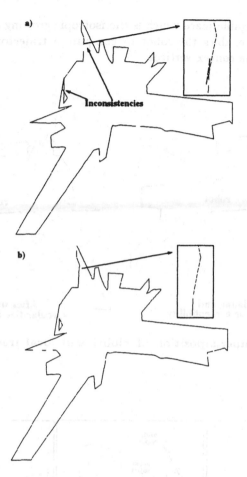

Fig. 8. Global Free-Space model inconsistencies and its updating

After local free-space building and global free-space updating, the respective models are exactly superimposed (fig 9) and the operation described in section 4 may be applied.

6 Experimentations

We conducted a large number of experiments using the described system. The results shown in figure 10 illustrate the algorithm described in section 4.

The second experimentation (fig 11) shows a "curious" mobile robot that moves in its environment and, when detecting the appearance of a new object, moves to viewpoints from wich it can have a complete perception of it, while modelling also the surrounding environment. In this figure we have sometimes represented the navigable space which is the isotropic growing of the free-space. This representation enables the robot to compute a trajectory by building a visibility graph of the convex vertices.

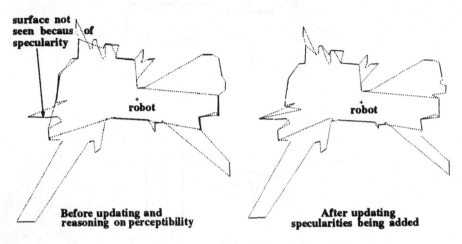

Fig. 9. Superimposition of global and local free-space models

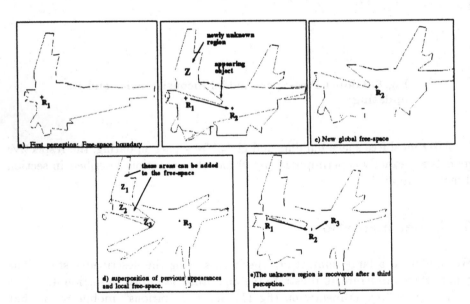

Fig. 10. Experimentation of the logical mechanism of free-space fusion

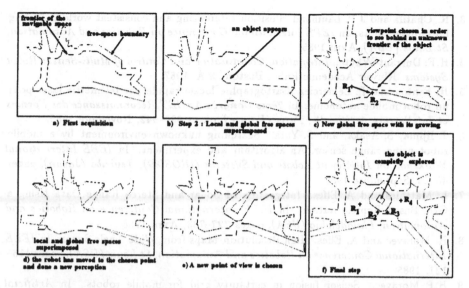

Fig. 11. The curious robot explores autonomously a new object that appeared within its known free-space by a reasoning on the perceptibility of the boundaries. The process takes into account sensor constraints.

7 Conclusion

We presented in this paper a complete approach for modelling the environment of a mobile robot, with an explicit algebraic representation of 2D object surfaces and of the free-space. The case of a non-static environment and non-perfect sensors were discussed. Such representations are important for environment understanding, path planning and reasoning for perception. The surfacic fusion process and the free-space incrementation operation are integrated in a global system called **Yaka**. **Yaka** uses these fusion methods for path planning and viewpoint selection for the mobile robot **Hilare**. Experimentations show the capabilities of this system to enable a mobile robot to move autonomously in its environment.

Acknowledgements. The authors wish to thank P. Gaborit and R. Alami for their important contributions to this work.

References

1. N. Ayache and O. Faugeras. Maintaining representations of the environment of a mobile robot. In *Robotics Research : The Fourth International Symposium, Santa Cruz (USA)*, Août 1987.
2. R.A. Brooks. Aspects of mobile robot visual map making. In *IEEE International Conference on Robotics and Automation, St Louis (USA)*, Mars 1985.

3. R. Chatila and J.P. Laumond. Position referencing and consistent world modeling for mobile robots. In *IEEE International Conference on Robotics and Automation, St Louis (USA)*, Avril 1985.

4. H.F. Durrant-Whyte. *Integration, Coordination and Control of Multi-Sensor Robot Systems*. Kluwer Academic Publ., Boston, MA, 1987.

5. H. Fanton and P. Lemercier. Cartographie locale rapide par télémètrie laser pour un robot mobile autonome. In *7ème Congrès AFCET Reconnaissance des Formes et Intelligence Artificielle, Paris (France)*, pages 933–942, Novembre 1989.

6. J. Iijima, S. Asaka, and S. Yuta. Searching unknown environment by a mobile robot using range sensor. An algorithm and experiment. In *IEEE International Workshop on Intelligent Robots and Systems (IROS'89), Tsukuba (Japan)*, pages 46–53, Septembre 1989.

7. L. Matthies and A. Elfes. Integration of Sonar and Stereo Range Data Using a Grid-Based Representation. In *IEEE International Conference on Robotics and Automation, Philadelphia (USA)*, pages 727–733, Avril 1988.

8. H. Moravec and A. Elfes. High Resolution Maps from Wide Angle Sonar. In *IEEE International Conference on Robotics and Automation, St Louis (USA)*, pages 116–121, 1985.

9. H.P Moravec. Sensor fusion in certainty grid for mobile robots. In *Artificial Intelligence Magazine*, pages 61–74, 1988.

10. P. Moutarlier. *Modélisation autonome de l'environnement d'un robot mobile*. PhD thesis, Thèse de l'Université Paul Sabatier, Toulouse (France), (Forthcoming) 1991.

11. P. Moutarlier and R. G. Chatila. An Experimental System for Incremental Environment Modelling by an Autonomous Mobile Robot. In *Proc. ISER, Montreal*, Juin 1989.

12. P. Moutarlier and R. G. Chatila. Stochastic Multisensory Data Fusion for Mobile Robot Location and Environment Modelling. In *Proc. International Symposium on Robotics Research, Tokyo*, 1989.

13. N. Nilsson. A mobile automaton: an application of artificial intelligence techniques. In *proc. of the first IJCAI*, pages 509–520, 1969.

14. R.C. Smith, M. Self, and P. Cheeseman. Estimating uncertain spatial relationships in robotics. In *2nd workshop on uncertainty in Artificial Intelligence. Philadelphia*, Aug. 1986.

Numerical Tools
for
Visual Perception

Matching 3-D Smooth Surfaces with their 2-D Projections using 3-D Distance Maps

Stéphane Lavallée 1, Richard Szeliski 2, Lionel Brunie 1

1 TIMB, IMAG
Faculté de Médecine de Grenoble
38700 La Tronche, France

2 Digital Equipment Corporation, Cambridge Research Lab
One Kendall Square, Bldg. 700, Cambridge, MA 02139

Abstract. The matching of 3-D anatomical surfaces to 2-D X-ray projections is an important problem in Computer and Robot Assisted Surgery. We present a new method for retrieving the rigid body transformation that describes this match. Our method performs a least squares minimization of the distance between the camera-contour projection lines and the surface. To correctly deal with projection lines that penetrate the surface, we minimize the square of the minimum signed distance along each line (distances inside the object are negative). To quickly and accurately compute distances to the surface, we represent the precomputed distance map using an octree spline whose resolution increases near the surface. The octree allows us to quickly find the minimum distance along each line using best first search. We present experimental results of 3-D surface to 3-D projection matching, and also show how our method extends to 3-D to 3-D surface matching.

Introduction

Matching a 3-D surface with its projections is a classical problem in computer vision. Our goal is to estimate the pose of an object from its projections, i.e., given an object in a coordinate system R_s [$_{obj}$] and given one or more contours of image projections of this object in a camera system R_s [$_{cam}$], the problem is to estimate the 6-component vector v that describes the rigid body transformation T_v between R_s [$_{obj}$] and R_s [$_{cam}$]. Although this problem can theoretically be solved for a single projection in practice, no solution based algorithms are necessary to have a sufficiently accurate estimate.

While the method we will present here has many applications, our primary interest is in solving multi-modality medical imaging matching problems. An important research topic in Computer and Robot Assisted Surgery is how to match 3-D images such as MRI (Magnetic Resonance Imaging) or CT scans (Computed Tomography) with X-ray projections [1]. Usually, the former images are pre-operative, while the latter are intra-operative. Matching the reference system [1] to associated with MRI or CT with the reference system [2] associated with X-ray projections enables one to perform robot assisted interventions based on pre-operative simulations performed on the 3-D images [2]. To achieve

Matching 3-D Smooth Surfaces with their 2-D Projections using 3-D Distance Maps

Stéphane Lavallée 1, Richard Szeliski 2, Lionel Brunie 1

1 TIMB - IMAG
Faculté de Médecine de Grenoble
38 700 La Tronche, France
2 Digital Equipment Corporation, Cambridge Research Lab
One Kendall Square, Bldg. 700, Cambridge, MA 02139

Abstract. The matching of 3-D anatomical surfaces to 2-D X-ray projections is an important problem in Computer and Robot Assisted Surgery. We present a new method for determining the rigid body transformation that describes this match. Our method performs a least squares minimization of the distance between the camera–contour projection lines and the surface. To correctly deal with projection lines that penetrate the surface, we minimize the square of the minimum *signed distance* along each line (distances inside the object are negative). To quickly and accurately compute distances to the surface, we represent the precomputed distance map using an *octree spline* whose resolution increases near the surface. The octree allows us to quickly find the minimum distance along each line using best-first search. We present experimental results of 3-D surface to 2-D projection matching, and also show how our method works for 3-D to 3-D surface matching.

1 Introduction

Matching a 3-D surface with its projections is a classical problem in computer vision. Our goal is to estimate the pose of an object from its projections, i.e., given an object in a coordinate system Ref_{3D} and given one or more contours of image projections of this object in a coordinate system Ref_{sensor}, the problem is to estimate the 6-component vector p that defines the rigid body transformation T between Ref_{sensor} and Ref_{3D}. Although this problem can theoretically be solved for a single projection, in practice two or more projections are necessary to achieve a sufficiently accurate estimate.

While the method we will describe has many applications, our primary interest is in solving multi-modality medical imaging matching problems. An important research topic in Computer and Robot Assisted Surgery is how to match 3-D images such as MRI (Magnetic Resonance Imaging) or CT scans (Computed Tomography) with X-ray projections [1]. Usually, the former images are pre-operative, while the latter are intra-operative. Matching the reference system Ref_{3D} associated with MRI or CT with the reference system Ref_{sensor} associated with X-ray projections enables us to perform robot assisted interventions based on pre-operative simulations performed on the 3-D images [2]. To achieve this objective, we first segment some reference anatomical structures both in 3-D in Ref_{3D} and in 2-D in Ref_{sensor} (the 2-D and 3-D segmentation problems are not addressed in this paper since they constitute their own full research topics [3, 4, 5]). After the segmentation step, we match a 3-D smooth surface (such as a vertebra) with two or more contours extracted from its X-ray projections. In other applications, the 3-D surface can be an *a priori* model of the object and/or the projections can be video camera images.

For our application, the main requirements are: (1) to perform the matching process for arbitrary free-form smooth surfaces; (2) to achieve the best accuracy possible for the 6 rigid body transformation parameters; (3) to compute an estimate of the uncertainties in the 6 parameters; and (4) to perform the matching process in a reasonable time. These requirements lead us to introduce the new formulation that we present in section 3 (a review of previous work in the general field of 3-D / 2-D matching is presented in section 2). Our new formulation requires fast and accurate computation of line to surface distances, which we solve using 3-D distance maps. Several distance maps are described in section 4. Of these, the fastest and most accurate computation is achieved using *octree splines* [6]. The fast computation of line to surface distances is described in section 5. Section 6 describes our method for estimating the 6 parameters p of the transformation T using nonlinear least squares minimization. Section 7 presents some experimental results. In section 8, we show how our method can be applied with few modifications to the 3-D / 3-D matching problem. Finally, the algorithm, other extensions of the method, and future improvements are discussed in section 9.

2 Previous work

Inferring 3-D objects from several images has been one of the classic research problems in computer vision. If a large set of images is used, *reconstruction* of 3-D objects is often possible with good accuracy. Computed Tomography solves this problem when the images are X-ray projections. Several different techniques can be used with video images (e.g., constructing 3-D mosaics from planar surfaces [7], constructing bounding volumes from silhouettes using an octree representation [8], or inferring surfaces from optical flow measurements [9]). When only a few images are used, *a priori* knowledge must be introduced to compensate for the lack of information. For example researchers have assumed that objects are described by points [10], line segments [11], planar primitives [12], or curves [13]. Using symmetry hypotheses has also been proved to be effective [14], even when the object does not exactly satisfy these hypotheses [15].

In some cases, the reconstruction of objects using models is reduced to the *recognition* of objects (e.g., methods looking for *invariants* in images have been widely studied [16, 17] or to the *positioning* of objects (this is the case when a full 3-D model of the object is known)[18]. In this paper, we address this latter problem. We assume that a complete representation of the 3-D surface of the object is known, but do not assume anything about the existence of edges or particular points or curves on the object.

Kriegman and Ponce have recently presented a detailed study of this matching problem [19]. They report that this problem was previously unsolved mainly because the contour generators of smooth surfaces (i.e., the curves that belong to the surface and that form the image contours) are not viewpoint-independent. Their paper presents a novel solution to this problem based on an algebraic approach (elimination theory) for surfaces represented by rational parametric patches. However, their solution works only if the surface is defined with a single rational patch. Since most free-form surfaces are represented by many patches, the problem of assigning 3-D patches to points on the image contours remains to be solved. Our approach is closer to classical *chamfer matching*, which iteratively minimizes the 2-D distance between features extracted from the image and projections of 3-D features selected from the 3-D model [20]. Instead of matching features, however, our method computes distances between projection lines and surfaces in 3-D.

3 Problem Formulation

In this section, we first introduce the representation and notation chosen for the rigid body transformations. We then briefly underline the importance of sensor calibration and present our formulation of the matching problem as the minimization of an energy.

3.1 Transformation and parameters

Using the notation defined in section 1, the problem is to estimate the transformation T between Ref_{sensor} and Ref_{3D}. T can be defined by a translation vector $t = (T_x, T_y, T_z)^t$ and a 3×3 rotation matrix R. Several representations can be used for R (e.g., Euler angles, quaternions, or rotation vectors [21]). We have chosen to use Euler angles $(\phi, \theta, \psi)^t$, where R is constructed from 3 rotations around x, y, z with angles ϕ, θ, ψ,

$$R = \begin{pmatrix} \cos\psi\cos\theta & -\sin\psi\cos\theta & \sin\theta \\ \cos\psi\sin\theta\sin\phi + \sin\psi\cos\phi & -\sin\psi\sin\theta\sin\phi + \cos\psi\cos\phi & -\cos\theta\sin\phi \\ -\cos\psi\sin\theta\cos\phi + \sin\psi\sin\phi & \sin\psi\sin\theta\cos\phi + \cos\psi\sin\phi & \cos\theta\cos\phi \end{pmatrix}. \tag{1}$$

The Euler representation is valid only if some restrictions on Euler angles are applied [21]. In our case, the Euler angles remain in a single domain during the convergence because the initial estimates are not too far away from the solution.

If we gather the 6 parameters of the transformation T into a 6-component vector

$$p = (T_x \ T_y \ T_z \ \phi \ \theta \ \psi)^t, \tag{2}$$

and use homogeneous coordinates, $T(p)$ can be represented by a single 4×4 matrix

$$T(p) = \begin{pmatrix} R & t \\ 0 & 1 \end{pmatrix}. \tag{3}$$

The relation between a point $q = (x\ y\ z\ 1)^t$ (or vector $v = (x\ y\ z\ 0)^t$) in Ref_{sensor} and the corresponding point $r = (X\ Y\ Z\ 1)^t$ (or a vector $u = (X\ Y\ Z\ 0)^t$) in Ref_{3D} can be written as

$$r = T(p)\, q \quad \text{or} \quad u = T(p)\, v. \tag{4}$$

3.2 Camera calibration

Because one of our requirements is good accuracy in the recovered parameters, accurate sensor calibration is very important [22]. We cannot use approximations such as orthography or weak perspective (also called scaled orthography). Moreover, we must model any image distortions that deviate from the perspective model. In order to take such *local* distortions into account with good accuracy, we have extended the 2-planes method [23, 24] to the *N-planes spline method* [25]. The result of these calibration processes is a function $C(\mathcal{P}_i) = L_i$ that associates every pixel $\mathcal{P}_i$ of each projection with a 3-D line whose representation is known in Ref_{sensor}. Each line $L_i = (q_i, v_i)$ is given by a point q_i and a vector v_i. In principle, we obtain a function C for each one of the N_P projections, but in practice, we can gather all of these N_P functions into one general function valid for any pixel of any projection. Note that we can use the above sensor model for both video cameras and X-ray projection systems.

3.3 Energy minimization

Given N_P projection contours of an object S, we first select a set of M_P pixels $\{\mathcal{P}_i, i = 1 \ldots M_P\}$, that belong to the N_P contours. This means that we do not need to perfectly segment the projection contours of S; we can in fact extract only some pixels $\mathcal{P}_i$ that lie on the contours with a high certainty. Each pixel $\mathcal{P}_i$ corresponds to a 3-D line L_i given in Ref_{sensor} through the function C defined above. As it is illustrated in figure 1, the transformation by $T(p)$ of each line $L_i = (q_i, v_i)$ in Ref_{sensor} into a line $l_i(p)$ in Ref_{3D} is given by

$$l_i(p) = (T(p)\, q_i, T(p)\, v_i) = (r_i(p), u_i(p)). \tag{5}$$

To compute the goodness of fit between a surface S in Ref_{3D} and its projections in Ref_{sensor}, we first define the 3-D *unsigned distance* between a point r and the surface S, $d_H(r, S)$, as the minimum Euclidean distance between r and all the points of S. We define the *signed distance*, $\tilde{d}(r, S)$, between a point r and the surface S as $d_H(r, S)$ if r is exterior to S, and as the negative of this distance if r is interior to S. Finally, we define a signed distance between a line $l_i(p)$ represented in Ref_{3D} and the surface S, $\tilde{d}_l(l_i(p), S)$, as the minimum of $\tilde{d}(r, S)$ over all the points that belong to $l_i(p)$,

$$\tilde{d}_l(l_i(p), S) \equiv \min_\lambda \tilde{d}(r_i(p) + \lambda u_i(p), S). \tag{6}$$

Note that whether the line $l_i(p)$ passes through the surface or not, as we look for the minimum value of the signed distance along $l_i(p)$, the absolute value of $\tilde{d}_l(l_i(p), S)$ indicates whether the line $l_i(p)$ is *close* and *tangent* to S (Figure 1).

When the correct matching between the contour projections and the surface S is reached, every line $l_i(p)$ is tangent to S, and every signed distance $\tilde{d}_l(l_i(p), S)$ is therefore zero. This leads us to formulate the matching problem as a least squares minimization of the energy or error function $E(p)$

$$E(p) = \sum_{i=1}^{M_P} \frac{1}{\sigma_i^2} [\tilde{d}_l(l_i(p), S)]^2, \qquad (7)$$

where σ_i^2 is the variance of the noise of the measurement $\tilde{d}_l(l_i(p), S)$ (see section 6). $E(p)$ is the weighted sum of squares of the signed distances between the projection lines $l_i(p)$ and the surface S. Given an initial estimate $p = p_0$ of the 3-D/2-

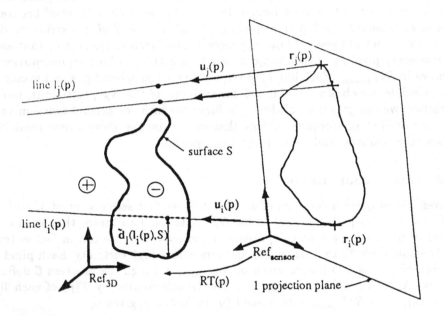

Fig. 1. Projection line to surface distance computation

D transformation parameters, a nonlinear least squares iterative minimization of the energy (or error function) $E(p)$ is performed using the Levenberg-Marquardt algorithm (see section 6). The term *energy* is used by analogy to potential field methods for obstacle avoidance in robotics [26]. Indeed, this minimization process corresponds to the evolving motion of an object (the set of all projection lines) in a potential field (the distance map $\tilde{d}$).

4 Fast Distance Computation Using Octree Splines

The method described in the previous section involves a fast computation of the distances $\tilde{d}$ and $\tilde{d}_l$. If the surface S is discretized in n^2 points, the computation of the distance $\tilde{d}$ is a $O(n^2)$ process. Similarly, if a line $l_i(p)$ is discretized in m points, the computation of the distance $\tilde{d}_l$ is a $O(mn^2)$ process. To speed up this process, we precompute a 3-D *distance map*, which is a function that gives the signed minimum distance to S from any point q inside a bounding volume V that encloses S. The next few paragraphs describe several 3-D distance maps that we have studied and implemented. In each of the following methods, the representation chosen of the surface is a set of n^2 points $s_i, i = 1 \ldots n^2$.

4.1 Uniform 3-D Distance Maps

The first representation that we studied and implemented was a uniform 3-D distance map. At each point q on a regular grid of N^3 points that describe V, the distance $\tilde{d}(q, S)$ is computed and stored. First, we look for the unsigned distance $d_H(q, S)$ between q and S by computing exhaustively the minimum value of the distance between q and all points s_i.

$$d_H(q, S) = \min_{s_i} d(q, s_i) = d(q, s_{i_{min}}), \tag{8}$$

where d is the Euclidean distance. In order to set negative values inside the surface, two solutions have been studied:

1. **Using surface normal vectors:** If the surface normal vector $\hat{n}_i$ is known at each point s_i (this is the case when the initial surface representation is parametric) and if this normal vector can be oriented towards the exterior of the surface, then $\tilde{d}(q, S)$ is computed by

$$\tilde{d}(q, S) = sign(\hat{n}_{i_{min}} \cdot (q - s_{i_{min}}))d(q, s_{i_{min}}). \tag{9}$$

   Another possibility would be to use

$$\tilde{d}(q, S) = (\hat{n}_{i_{min}} \cdot (q - s_{i_{min}}))d(q, s_{i_{min}}). \tag{10}$$

   Both (9) and (10) can be used for surfaces that are not closed. Unfortunately, if the surface is not discretized finely enough or if the normal vectors are not accurate enough, this method leads to artifacts such as negative values far away from the surface.

2. **Spreading a sign from an exterior point:** In this method, $\tilde{d}$ is computed in two passes. In the first pass, each $\tilde{d}(q, S)$ is set to $-d_H(q, S)$ using (8). Then, a positive sign is spread from an initial exterior point. This recursive spreading of the positive sign from points to their neighbors is terminated at points close to the surface (using a threshold on the distance $d_H(q, S)$).

Once $\tilde{d}$ has been computed for all points on the regular grid, the distance $\tilde{d}$ can be computed at arbitrary points q inside the volume V by interpolating grid values. Both trilinear interpolation and tricubic interpolation with tensorial spline functions have been implemented.

In the preceding method, because computing the distance map has a $O(n^2 N^3)$ complexity, computation times rapidly become prohibitive. In order to reduce these computation times, we have used the classical *chamfer algorithm* [27, 28, 29] in 3 dimensions, which requires only two sweeps of V. Unfortunately, uniform distance maps are not well suited to quickly finding the minimum distance along a line $\bar{d}_l$ (see section 5). Moreover, achieving acceptable accuracy can be prohibitively expensive in terms of memory space (typically 256^3 nodes).

4.2 Octree Spline Distance Maps

In looking for an improved trade-off between memory space, accuracy, speed of computation, and speed of construction, we have developed a new kind of distance map which we call the *octree spline*. The intuitive idea behind this geometrical representation is to have more detailed information (i.e., more accuracy) near the surface than far away from it. We start with the classical octree representation associated with the surface S [30] and then extend it to represent a continuous 3-D function that approximates the signed Euclidean distance to the surface. This representation combines advantages of adaptive spline functions and hierarchical data structures. In the actual implementation, this 3-D function is a piecewise C^0 function, but it is intended to be extended to a piecewise C^2 function in next implementations, so this consideration justifies the use of the name *spline*.

As for the previously described distance maps, the input to the octree spline construction algorithm is a set of n^2 points s_i regularly spread on the surface S. The algorithm performs the following steps:

1. **Surface point octree construction:** First, the octree associated to the set of points s_i is built according to classic octree subdivision [30]. Starting from the initial cube V, each node that contains points (grey node) is recursively split into 8 sub-cubes until it contains no points (white node) or it has the maximal selected resolution (black node). For our applications, the octree typically has 6 to 8 levels, corresponding to a resolution of 1/64 to 1/256. At the end of this step, each node (grey, black or white) contains the list of surface points that are inside the node.

2. **Subdivision (refinement):** The octree previously computed may have large sized nodes near the surface, because no rules about subdivision near the surface have been introduced. To overcome this problem, we perform a further subdivision of the octree to ensure that two nodes which are neighbors along a face, edge, or corner differ in size by at most a factor of k_S (in practice, we choose $k_S = 2$). In order to compute the 26 neighbors of any node in the octree, we use the technique described in [31]. The resulting structure is called a *restricted octree* [32].

3. **Corner distance computation:** For each corner c of each terminal node (white or black), the distance $d(c, S)$ is computed according (8). The spatial organization of surface points created by the octree makes this process much faster since we can use an efficient search technique to find the minimum distance between c and all of the points s_i. For each corner $c = (U_c, V_c, W_c)$ of a node N, the octree is visited starting from N and recursively visiting the

parents and neighbors of N. A best value d_{best} for the minimum distance is maintained during this process. If a candidate node has central coordinates (U, V, W) and size s such that

$$|U-U_c| > d_{best}+\frac{s}{2} \quad \text{or} \quad |V-V_c| > d_{best}+\frac{s}{2} \quad \text{or} \quad |W-W_c| > d_{best}+\frac{s}{2}, \quad (11)$$

the points s_i lying inside this node are not tested since the minimum distance would be guaranteed to be greater than d_{best}.

4. **Signed distance computation:** In order to compute $\tilde{d}$ from d_H, the two different methods described for uniform distance maps have been implemented for octree splines. As in that previous case, we prefer the recursive spreading of a sign starting from an exterior point. Here, the recursive spreading stops when a black node is encountered (rather than when a low threshold on distances is passed). This works if the black nodes constitute a connected set of nodes.

5. **Crack elimination (continuity enforcement):** Because the signed distance is computed at any point q by an interpolation based on the 8 corner values of the terminal node that contains q (see below), discontinuities or *cracks* can appear if we simply interpolate the true corner distance values. Several solutions can be used to suppress the cracks [30]. We chose the following process for its simplicity: if a corner c of a node N_1 of size s_1 lies on a face or an edge of another node N_2 of size $s_2 > s_1$, then the distance value of the corner c is simply replaced by the distance computed at c by interpolation inside N_2. This method is applied to the octree with a top-down breadth-first traversal. Note that this process introduces only a small loss of accuracy since the octree had been previously *restricted* at step 2.

After the previous steps have been performed, $\tilde{d}(q, S)$ can be computed for any point q using a trilinear interpolation of the 8 corner values $\tilde{d}_{ijk}$ of the terminal node N that contains the point q (a classical binary search starting from the root is used to find N). If $(u, v, w) \in [0, 1] \times [0, 1] \times [0, 1])$ are the normalized coordinates of q in the cube N,

$$\tilde{d}(q, S) = \sum_{i=0}^{1}\sum_{j=0}^{1}\sum_{k=0}^{1} b_i(u)b_j(v)b_k(w)\tilde{d}_{ijk} \quad \text{with} \quad b_l(t) = \delta_l t+(1-\delta_l)(1-t). \quad (12)$$

We can compute the gradient $\nabla\tilde{d}(q, S)$ of the signed distance function by simply differentiating (12) with respect to u, v, and w. Because $\tilde{d}$ is only C^0, $\nabla\tilde{d}(q, S)$ is discontinuous on cube faces. However, these gradient discontinuities are relatively small and do not seem to affect the convergence of our iterative minimization algorithm.

One additional pre-processing step is required in order to speed up the line minimization algorithm presented in the next section:

6. **Lower bound estimation:** Compute for each node N of the octree a function B_N which is a lower bound on the function $\tilde{d}$ inside N,

$$\forall q \in N, \quad B_N(q) \le \tilde{d}(q, S). \quad (13)$$

A suitable choice for a bound function, with a good trade-off between simplicity and accuracy, is the linear approximation

$$B_N(q) = au + bv + cw + d. \tag{14}$$

Because trilinear interpolation (12) is used to compute $\tilde{d}_l(q, S)$, the inequality (13) can be solved recursively (bottom-up) for (a, b, c, d) using a modified simplex algorithm.

5 Line To Surface Distance Minimization

This section describes how the signed distance from a line to the surface $\tilde{d}_l(l_i(p), S)$ defined in (6) can be rapidly computed using the octree spline. One possible approach to computing the minimum expressed in (6), which works for any representation of the distance map, would be to use classical one-dimensional functional minimization. Because the function $f(\lambda) = \tilde{d}(r + \lambda u, S)$ may have several local minima, global search techniques such as simulated annealing may have to be used. Another alternative that works with a discretized distance map would be to exhaustively search all cells that the line intersects. For a uniform discretization of the distance map, this takes $(O(N))$ steps, which may be prohibitively slow.

Our approach is to take advantage of the octree spline to develop an optimized search technique that requires approximately $O(\log(N))$ steps (where N is the resolution of the octree). Because our octree representation of the distance map is tree-based, we could use a variety of tree-search techniques. We have chosen to use best-first search with lower bounds computed for each node (interior and terminal) of the octree spline. Best-first search requires a priority queue, which we implement using a heap. In our algorithm, a candidate line $l_i(p)$ is recursively split by the octree cube boundaries into smaller line segments $\Sigma = [a, b]$ until the line segment with minimum distance is found. During this process, the current minimum distance d_{best} is maintained, and an estimated lower bound on the distance

$$B_N(\Sigma) = \min(B_N(a), B_N(b)) \tag{15}$$

is computed for each new segment before it is pushed onto the priority queue. As segments are popped from the queue in order of smallest lower bound, they are either split into smaller segments (if the associated node is non-terminal), or tested to see if the minimum attainable distance

$$\tilde{d}_l(\Sigma, S) = \min(\tilde{d}(a, S), \tilde{d}(b, S)) \tag{16}$$

is lower than d_{best} (in which case a or b becomes the new minimum point). Note that while (16) is not the exact minimum distance over the line, the approximation is quite good, especially near the correct pose, where the line segments Σ are tangent to the surface S and the corresponding nodes N have the minimal size. The best-first search is terminated when the smallest lower bound on the priority queue is greater than d_{best}.

6 Least Squares Minimization

This section describes the nonlinear least squares minimization of the energy or error function $E(p)$ defined in (7). Least squares techniques work well when we have many uncorrelated noisy measurements with a normal (Gaussian) distribution[3]. To begin with, we will make this assumption, even though noise actually comes from calibration errors, 2-D and 3-D segmentation errors, the approximation of the Euclidean distance by octree spline distance maps, and non-rigid displacement of the surface between Ref_{3D} and Ref_{sensor}. Moreover, if we used unsigned distances d_H instead of $\tilde{d}$, the algorithm would still converge towards the correct solution in case no noise would exist. However, for real noisy data, as all lines $l_i(p)$ passing through the surface give a null error component $\tilde{d}_l(l_i(p), S)$, a bias would be introduced in the estimation of p.

To perform the nonlinear least squares minimization, we use the Levenberg-Marquardt algorithm because of its good convergence properties [33]. Merging(5), (6) and (7), we get

$$E(p) = \sum_{i=1}^{M_P} \frac{1}{\sigma_i^2}[e_i(p)]^2 = \sum_{i=1}^{M_P} \frac{1}{\sigma_i^2}[\min_\lambda \tilde{d}(T(p)q_i + \lambda T(p)v_i, S)]^2. \quad (17)$$

In order to compute the gradient and Hessian of $E(p)$, the Levenberg-Marquardt algorithm requires the first derivatives of each $e_i(p)$. For any component p_j of p, we obtain

$$\frac{\partial e_i(p)}{\partial p_j} = [\nabla \tilde{d}(T(p)[q_i + \lambda_{\min} v_i], S)] \cdot \left[\left(\frac{\partial T(p)}{\partial p_j}\right)(q_i + \lambda_{\min} v_i)\right], \quad (18)$$

where $\lambda_{\min}$ is the value of λ where the minimum is reached in (17). A surprising result is that (18) is valid (except in the degenerate case $\partial^2 \tilde{d}/\partial \lambda^2 = 0$) even though it ignores the variation of $e_i(p)$ with respect to λ. Thus, computing the 6-component gradient of $E(p)$ only requires computing the gradient of the octree spline distance $\tilde{d}$ (by differentiating (12)) and computing the 3 derivatives of $T(p)$ with respect to the 3 Euler angles (the 3 derivatives of $T(p)$ with respect to translational components are simple constants).

The end of the iterative minimization process is reached either when $E(p)$ is below a fixed threshold, when the difference between parameters $|p^{(k)} - p^{(k-1)}|$ at two successive iterations is below a fixed threshold, or when a maximum number of iterations is reached. At this point, we compute a robust estimate of the parameter p by throwing out the measurements where $e_i^2(p) \gg \sigma_i^2$ and performing some more iterations [34]. This process removes the influence of the *outliers* which are likely to occur in the automatic 2-D and 3-D segmentation processes (for instance, a partially superimposed object on X-ray projections can lead to false contours). The threshold for outlier rejection must be fixed according to application-specific knowledge or by experimentation. In our case,

---

[3] Under these assumptions, the least squares criterion is equivalent to maximum likelihood estimation.

we chose to set this threshold to 3σ where σ is a mean *a priori* standard deviation of the noise.

Using a gradient descent technique such as Levenberg-Marquardt we might expect that the minimization would fail because of local minima in the 6-dimensional parameter space. However, for the experiments we have conducted, false local minima were few and always far away from the solution. Since in our application we have a good initial estimate of the parameters, these other minima are unlikely to be reached.

Finally, at the end of the iterative minimization procedure, we estimate the uncertainty in the parameters. The covariance matrix $Cov(p)$ is computed by inverting the final Hessian matrix and eigenvalue analysis of $Cov(p)$ is performed [33]. The result is an ellipsoid of uncertainties in the 6-dimensional parameter space. The distribution of errors after minimization is also computed in order to check that it is Gaussian.

7 Experimental results

The octree spline construction and the matching algorithm have been implemented in C, using X Windows on a DECstation 5000/200. Computation times expressed below are given for this machine.

7.1 Graphical tools

As part of our implementation, we have developed a suite of real-time interactive 3-D graphical tools which help us to develop our algorithms, to visualize their performance, and to explore various parameter settings. Our graphical programming environment has been used to visualize both the octree spline construction process and the iterative matching process. We can selectively visualize any object at any time: 3-D surface, projection frames, projection curves, projection lines, and octree frame at different resolutions (Figures 2 to 4). The viewpoint can be interactively modified to perform rotations, translations or zooms of the scene, or set to coincide with one of the real conic projection viewpoints (Figure 5). 2-D slices of the octree spline can be computed in real-time, and the quadtree decomposition of the slice or a pseudocolored rendering of the signed distance function can $\tilde{d}$ be displayed (Figure 3). Individual components of the algorithms can be also tested. For example, the distance value $\tilde{d}$ can be interactively displayed for any point on a slice, or, for interactively selected lines, the minimum value $\tilde{d}_l$ can be computed in real-time, and the octree node where the minimum is reached can be displayed (Figure 2b).

7.2 Experiments

This section describes our experiments with 3-D/2-D matching. We have performed tests on both real anatomical surfaces (surface S_1, Figure 4) and on

228

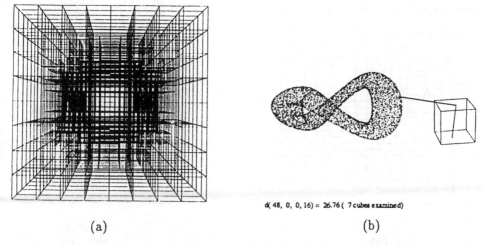

$d(48, 0, 0, 16) = 26.76 (\ 7\ cubes\ examined)$

(a)   (b)

Fig. 2. Octree frame and line to surface distance: (a) octree frame for surface S_2 (only 5 resolution levels are displayed), (b) line to surface distance computation: for an interactively selected line, the node where the minimum signed distance to the surface is found is displayed and the gradient vector at this point is drawn with a length equal to the minimum distance.

simulated surfaces (surface S_2, Figure 6). The projection curves of these surfaces were obtained by simulation in order to know the parameters p^* for which the correct pose is reached. Tests on real projections are under development. The different steps of our experiments are the following:

1. Simulate a transformation $T(p^*)$ applied to S and compute the N_P silhouettes of the transformed surface $S'(p^*)$ by projecting all the surface points
2. Extract the N_P contours of the silhouettes of $S'(p^*)$
3. Randomly select M_P pixels on the N_P contours (a percentage of all the contour pixels is chosen, typically 10%) and for each selected pixel compute the corresponding projection line according to the calibrated projection parameters (simulated conic projection)
4. Compute the octree spline $\tilde{d}$ associated with the original surface S
5. Starting from an initial estimate $p^{(0)}$, use the Levenberg-Marquardt algorithm to iteratively minimize the function $E(p)$ defined by (17), using the gradient expressions given in (18). At each iteration k, compute and display the difference $\Delta p = p^{(k)} - p^*$ between the current parameters $p^{(k)}$ and the

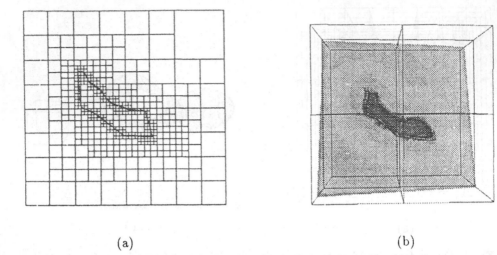

(a)   (b)

Fig. 3. 2D slice of the octree spline (surface S_1) showing the intersection between a plane interactively selected by the user and the octree spline: (a) slice of the octree frame and intersection curve between the plane and S_1, (b) slice of the octree spline in pseudo-color: shaded red (dark grey) for negative values inside S_1 and shaded green (light grey) for positive values outside.

simulated parameters p^*. Compute the error transformation

$$\Delta T^{(k)} = [T(p^{(k)})][T((p^*)]^{-1} \tag{19}$$

and extract the translation error

$$\Delta t^{(k)} = g - \Delta T^{(k)}g,$$

where g is the center of gravity of the surface points in Ref_{3D}. Extract $\Delta\alpha^{(k)}$, which is the angle of the rotation component R of $\Delta T^{(k)}$. The values of $\|\Delta t^{(k)}\|$ and $|\Delta\alpha^{(k)}|$ are displayed to monitor the convergence of the algorithm towards the optimal solution.

6. Perform robust estimation by removing the outliers and performing some more iterations.
7. Perform error analysis: compute the covariance matrix and the corresponding eigenvalues and eigenvectors; compute and display the error distribution.

Figures 4 to 7 show the results of some of the experiments that we have conducted. Figures 4 and 5 show the state of the iterative minimization algorithm for surface S_1 after 0, 2, and 6 iterations. Figure 4 shows the relative positions of the projections lines and the surface seen from a general viewpoint. Figure 5 shows the same state seen from the viewpoints of the two cameras. Figure 6 shows similar results for surface S_2. A plot of the algorithm convergence (energy, translation error, and rotation error) is shown in Figure 7.

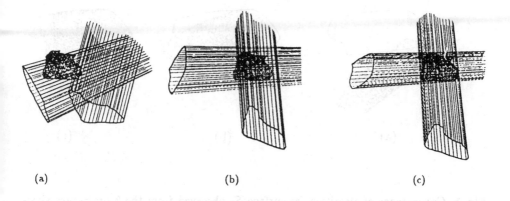

(a)   (b)   (c)

Fig. 4. Convergence of algorithm observed from a general viewpoint (surface S_1 is represented by a set of points). Two sets of projection lines evolve in the 3D potential field associated with the surface until each line is tangent to S_1: (a) initial configuration, (b) after 2 iterations, (c) after 6 iterations. For this case, the matching is performed in 1.8 s using 77 projection lines, in 0.9 s using 40 projection lines.

As we can see from these results, the algorithm very quickly finds the optimal match between the surface and its projections.

8 Matching 3-D objects

Our 3-D / 2-D matching algorithm has been extended to solve 3-D / 3-D matching problems. Here we look for the transformation $T(p)$ between a surface S known in Ref_{3D} and a set of M_P points q_i known in Ref_{sensor} (we make the assumption that most of the points q_i match to the surface). Our target application is to match pre-operative images such as CT and MRI (in which a reference surface S has been segmented) with intra-operative range images or ultrasound images (in which clouds of points that belong to the surface S can be acquired). A specific method that only solves this problem for convex surfaces (such as the head) is described in [35]. Methods for matching 2 surfaces (not a surface with a set of points) are also reported in [36] (extraction and matching

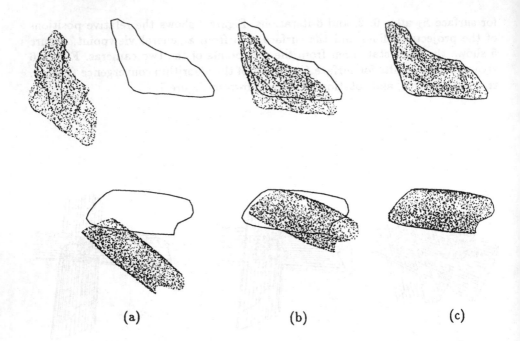

(a)   (b)   (c)

Fig. 5. Convergence of algorithm for surface S_1 observed from the 2 projection view-points. The external contours of the projected surface end up fitting the real contours: (a) initial configuration, (b) after 2 iterations, (c) after 6 iterations.

of singularities) and in [37](extraction and matching of radial contours). Global matching methods have been presented when surfaces can be matched at all points [38](extraction and matching of axes of inertia).

The formulation we introduced previously can be used for the 3-D / 3-D matching problem. The error measure (17) is now replaced by

$$E(p) = \sum_{i=1}^{M_P} \frac{1}{\sigma_i^2} [e_i(p)]^2 = \sum_{i=1}^{M_P} \frac{1}{\sigma_i^2} [\tilde{d}(T(p)\, q_i, S)]^2 .. \tag{20}$$

The gradient of $E(p)$ is computed using

$$\frac{\partial e_i(p)}{\partial p_j} = [\nabla \tilde{d}(T(p)\, q_i, S)] \cdot \left[\left(\frac{\partial T(p))}{\partial p_j} \right) (q_i) \right]. \tag{21}$$

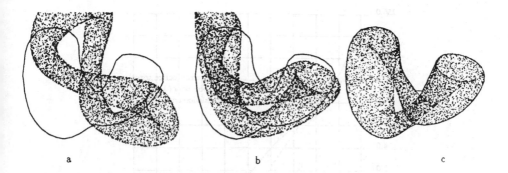

a   b   c

Fig. 6. Convergence of algorithm for surface S_2 (twisted torus) observed from a general viewpoint. For this case, the matching is performed in 5 s using 135 projection lines, in 1 s using 15 projection lines.

(a) initial configuration, $E(p^{(0)})/M_P = 63.87$, $\|\Delta t^{(0)}\| = 44.10mm$, $|\Delta\alpha^{(0)}| = 48.25°$,
(b) after 2 iterations, $E(p^{(2)})/M_P = 13.33$, $\|\Delta t^{(2)}\| = 10.72mm$, $|\Delta\alpha^{(2)}| = 14.75°$,
(c) after 10 iterations. $E(p^{(10)})/M_P = 0.91$, $\|\Delta t^{(10)}\| = 0.21mm$, $|\Delta\alpha^{(10)}| = 0.16°$.

As in 3-D / 2-D matching, the distance map $\tilde{d}$ is also built and represented using an octree spline. Actually, since the distance need not be signed (negative inside/positive outside), the octree spline can be built for non-closed surfaces and without knowledge of surface normal vectors. However, using a signed distance is still preferable since the zeros of the 3-D function $\tilde{d}$ coincide more accurately to the 3-D surface than the minima of the unsigned distance function [6].

Experiments have also been conducted to test this method. For an anatomical surface S_3 (surface of a vertebra, Figure 8), we compute the associated octree spline and select a subsampled set of M_P surface points q_i. A transformation $RT(p^*)$ is applied to this set of points and the robust estimation of the parameters is performed as for the 3-D / 2-D matching by minimizing the energy $E(p)$ expressed in (20). The results of this minimization for surface S_3 are shown in Figure 8.

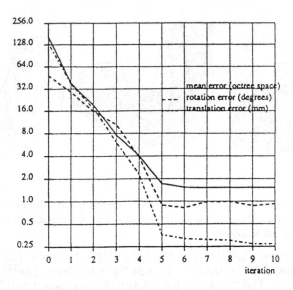

Fig. 7. Typical convergence curves of the matching algorithm (surface S_1) showing the mean error $E(p^{(k)})/M_P$ (in voxels), translation error $\|\Delta t^{(k)}\|$ (in mm), and rotation error $|\Delta\alpha^{(k)}|$ (in degrees) as a function of iteration number k.

9 Discussion

9.1 Matching method

In this paper, we have developed an algorithm for 3-D surface to 2-D contour matching that involves minimizing the 3-D distance between the projection lines and the surface. This is not the first formulation that we tried. We first implemented a method that iteratively projects the surface S transformed by $T(p)$ and minimizes the distance between the contours C_S of these simulated projections and the contours C_R segmented on the real input images. What are the advantages of our new method with respect to this alternative approach?

First, computing the silhouette contours of the surface projection is more time consuming than looking for the minima of the octree spline along a set of projection lines. In fact, when projection models have to take local distortions into account, a fast way of computing the silhouette contours of a projection of S is to use a function that tests for any pixel if the corresponding projection line intersects the surface (this is equivalent to our minima search along projection lines). However, this function must be computed at each iteration for all pixels on the contour and for all of their neighbors [39]. In our method, only a few pixels on the contours are used.

Second, the distance between two arbitrary 2-D contours that may only have partial overlap can be difficult to define and to minimize. It may be possible to use some *global* characteristics such as angles between inertia moments of contours and to minimize this distance [39], but this relies on a full 2-D segmentation of real contour projections, which can be difficult, especially for X-ray

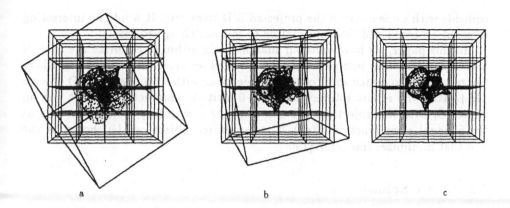

Fig. 8. Convergence of 3-D/3-D matching algorithm for surface S_3 (vertebra) segmented from a 3D CT image. For this case, the matching is performed in 2 s using 130 data points.

(a) initial configuration, $E(p^{(0)})/M_P = 113.47$, $\|\Delta t^{(0)}\| = 125.23mm$, $|\Delta\alpha^{(0)}| = 48.25°$,
(b) after 2 iterations, $E(p^{(2)})/M_P = 38.58$, $\|\Delta t^{(2)}\| = 25.97mm$, $|\Delta\alpha^{(2)}| = 20.53°$,
(c) after 6 iterations. $E(p^{(6)})/M_P = 4.20$, $\|\Delta t^{(6)}\| = 0.75mm$, $|\Delta\alpha^{(6)}| = 0.32°$.

projections. A sum of local distances is thus more appropriate. Since fast computation is a requirement, a 2-D distance map must be computed for each real contour projection [20]. However, real contours can be difficult to obtain reliably and 2-D segmentation errors will be propagated to the distance map, which may lead to unpredictable results. Moreover, careful examination shows that this minimization problem cannot be solved using least squares techniques such as the Levenberg-Marquardt algorithm. Classical minimization of multivariate functions must be used instead [33]. Moreover in this technique, gradient components cannot be computed analytically (while in our new method, we can use (18) and (21)). Approximation of gradient by finite differences is thus required, which means computing 6 more projection contours at each iteration and weakening the convergence due to inaccurate gradients.

A potential weakness of our method is that we use only the contour information in the projected images. In X-ray images, the grey levels give us information about the integral of the density along the corresponding projection line. The octree spline representation used in this paper could be extended to model such continuous densities, and would help in rapidly computing the integrals. Our method also ignores the discontinuities in the projected contours, which must coincide with some event in the projected 3-D scene [40]. It would be interesting to see if we could add this information to our matching process.

In this paper, we have used our matching algorithm to estimate the pose of a known object. Our method could also be used for *recognition* problems, where the purpose is to match some contour projections with a finite set of 3-D objects $\{O_i\}$. After the robust estimation of the 6 attitude parameters p^* that link an object O_i with its projections, we can compute a *matching score* by looking at the residual error function $E(p^*)$. The recognized object can be chosen as the one that minimizes this score [19].

9.2 Octree Splines

The octree spline representation introduced in this paper has proven to be very effective at solving our matching problem. We are currently studying how to extend the octree spline and how to use it as a general modeling tool for a variety of applications.

First, we are studying how to construct octree splines from inputs more sophisticated than unordered point lists. Examples of such input surface representations include parametric splines [4], arbitrary triangulations, and level crossings in volumetric (voxel) data. Indeed, in the actual implementation, the points used to build the octree-spline have to be more or less regularly spread on the surface, in order to give an homogeneous accuracy. Using higher-levels representations (triangles or patches) as input of the octree-spline overcomes this constraint. Second, we are examining how to speed up the computation of the distance map $\tilde{d}$ at the corners of the octree[4], using generalizations of chamfer distances to octrees as well as approximate techniques based on fast N-body solvers [41].

Finally, we are studying methods for extending the continuity of the octree spline from C^0 to C^1 and higher order continuities. The difficulty here is to merge overlapping spline basis functions with the octree representation. Compared to these other representations, octree splines have the potential to be less space consuming and to permit the rapid evaluation of certain geometric operations. Possible novel applications for the octree-spline include: obstacle avoidance in robotics using potential field methods; fast computation of intersections curves between two complex 3-D surfaces; and surface interpolation of arbitrary point sets, using the zero-crossings of a C^k octree spline to implicitly represent the surface.

10 Conclusions

In this paper, we have presented a new method for estimating the pose of a 3-D surface by matching it with a set of 2-D projections. We have demonstrated

---

[4] This operation currently takes about 15 minutes on a DECstation 5000/200 for a 7-levels octree with 100,000 points

the performance of our approach on a number of examples, and shown how it can be extended to solve 3-D / 3-D matching problems. The four requirements presented in the introduction have all been met. First, the matching process works for any free-form smooth surface, since no *a priori* assumptions (symmetry hypotheses, algebraic forms, or special curves) are used. Second, we achieve the best accuracy possible for the estimation of the 6 parameters in p, because the octree spline representation we use approximates the true 3-D Euclidean distance with an error smaller than the segmentation errors in the input data. Third, we provide an estimate of the uncertainties of the 6 parameters, using least squares estimation with the Levenberg-Marquardt algorithm to compute these uncertainties. Fourth, we perform the matching process in rapidly, using the octree spline to compute a signed distance from a line to a surface very quickly. Our next goal is to obtain real-time performance so as to integrate our algorithm into clinical robot-assisted surgical applications.

References

1. S. Lavallee, L. Brunie, B. Mazier, and P. Cinquin. Matching of medical images for computer and robot assisted surgery. In *IEEE EMBS Conference*, Orlando, Florida, November 1991.
2. S. Lavallee and P. Cinquin. Computer assisted medical interventions. In K.H. Hohne, editor, *NATO ARW, Vol F60, 3D Imaging in Medicine*, pages 301–312, Berlin, June 1990. Springer-Verlag.
3. S. Leitner, I. Marque, S. Lavallee, and P. Cinquin. Dynamic segmentation : finding the edge with spline snakes. In P.J. Laurent, editor, *International Conference on Curves and Surfaces*, Chamonix, 1991. Academic Press.
4. I. Marque. *Segmentation d'Images Medicales Tridimensionnelles Basee sur une Modelisation Continue du Volume*. PhD thesis, Grenoble University, France, December 1990.
5. O. Monga, R. Deriche, G. Malandrain, and J. P. Cocquerez. Recursive filtering and edge closing : Two primary tools for 3D edge detection. In *First European Conference on Computer Vision (ECCV'90)*, pages 56–65, Antibes, France, April 1990. Springer-Verlag.
6. R. Szeliski and S. Lavallée. Octree splines. (in preparation) 1991.
7. M. Herman, T. Kanade, and S. Kuroe. Incremental acquisition of a three-dimensional scene model from images. *IEEE Transactions on Pattern Analysis and Machine Intelligence*, PAMI-6(3):331–340, May 1984.
8. R. Szeliski. Real-time octree generation from rotating objects. Technical Report 90/12, Digital Equipment Corporation, Cambridge Research Lab, December 1990.
9. R. Szeliski. Shape from rotation. In *IEEE Computer Society Conference on Computer Vision and Pattern Recognition (CVPR'91)*, Maui, Hawaii, June 1991. IEEE Computer Society Press.
10. M.A. Fischler and R.C. Bolles. Random sample consensus: A paradigm for model fitting with applications to image analysis and automated cartography. *Communications of the ACM*, 24(6):381–395, June 1981.
11. R. Horaud. New methods for matching 3D objects with single perspective views. *IEEE PAMI*, 9(3):401–412, 1987.

12. D. Cyganski and Orr J. A. Application of tensor theory to object recognition and orientation determination. *IEEE PAMI*, 7(6):662–673, 1985.

13. D. G. Lowe. *Perceptual Organization and Visual Recognition*. Kluwer Academic Publishers, Boston, Massachusetts, 1985.

14. M. Dhome, J. T. Lapreste, G. Rives, and M. Richetin. Spatial localization of modelled objects of revolution in monocular pespective vision. In *First European Conference on Computer Vision (ECCV'90)*, pages 475–485, Antibes, France, April 1990. Springer-Verlag.

15. D. Terzopoulos, A. Witkin, and M. Kass. Constraints on deformable models: Recovering 3D shape and nonrigid motion. *Artificial Intelligence*, 36:91–123, 1988.

16. Y. Lamdan, J. T. Schwartz, and H. J. Wolfson. Object recognition by affine invariant matching. In *IEEE Computer Society Conference on Computer Vision and Pattern Recognition (CVPR'88)*, pages 335–344, Ann Arbor, Michigan, June 1988. IEEE Computer Society Press.

17. D. Forsyth, J. L. Mundy, A. Zisserman, and C. M. Brown. Projectively invariant representations using implicit algebraic curves. In *First European Conference on Computer Vision (ECCV'90)*, pages 427–436, Antibes, France, April 1990. Springer-Verlag.

18. M. Dhome, M. Richetin, J. T. Lapreste, and G. Rives. Determination of the attitude of 3d objects from a single perspective view. *IEEE PAMI*, 11(12):1265–1278, 1989.

19. D. J. Kriegman and J. Ponce. On recognizing and positioning curved 3-D objects from image contours. *IEEE Transactions on Pattern Analysis and Machine Intelligence*, PAMI-12(12):1127–1137, December 1990.

20. H. G. Barrow, J. M. Tenenbaum, R. C. Bolles, and H. C. Wolf. Parameteric correspondence and chamfer matching: Two new techniques for image matching. In *Fifth International Joint Conference on Artificial Intelligence (IJCAI-77)*, pages 659–663, Cambridge, Massachusetts, August 1977.

21. N. Ayache. *Artificial Vision for Mobile Robots: Stereo Vision and Multisensory Perception*. MIT Press, Cambridge, Massachusetts, 1991.

22. R. Y. Tsai. Synopsis of recent progress on camera calibration for 3D machine vision. In *The Robotics Review*, pages 147–160. MIT Press, 1989.

23. H. A. Martins, J. R. Birk, and R. B. Kelley. Camera models based on data from two calibration planes. *Computer Vision, Graphics, and Image Processing*, 17:173–179, 1981.

24. K. D. Gremban, C. E. Thorpe, and T. Kanade. Geometric camera calibration using systems of linear equations. In *IEEE International Conference on Robotics and Automation*, pages 562–567, Philadelphia, Pennsylvania, April 1988. IEEE Computer Society Press.

25. G. Champleboux. *Utilisation des fonctions splines a la mise au point d'un capteur tridimensionnel sans contact : application a la ponction assistee par ordinateur*. PhD thesis, Grenoble University, July 1991.

26. J.C. Latombe. *Robot Motion Planning*. Kluwer Academic, Norwell,MA, 1991.

27. P.-E. Danielson. Euclidean distance mapping. *Computer Graphics and Image Processing*, 14:227–248, 1980.

28. G. Borgefors. Distance transformations in arbitrary dimensions. *Computer Vision, Graphics, and Image Processing*, 27:321–345, 1984.

29. G. Borgefors. Distance transformations in digital images. *Computer Vision, Graphics, and Image Processing*, 34:344–371, 1986.

30. H. Samet. *The Design and Analysis of Spatial Data Structures.* Addison-Wesley, Reading, Massachusetts, 1989.

31. G. Garcia. *Contribution a la modelisation d'objets et a la detection de collisions en robotique a l'aide d'arbres octaux.* PhD thesis, Nantes University, september 1989.

32. B.V. Herzen and A.H. Barr. Accurate triangulations of deformed, intersecting surfaces. *Computer Graphics,* 21(4):103–110, 1987.

33. W. H. Press, B. P. Flannery, S. A. Teukolsky, and W. T. Vetterling. *Numerical Recipes: The Art of Scientific Computing.* Cambridge University Press, Cambridge, England, 1986.

34. P. J. Huber. *Robust Statistics.* John Wiley & Sons, New York, New York, 1981.

35. C. A. Pelizzari, G. T. Y. Chen, D. R. Spelbring, R. R. Weichselbaum, and C. T. Chen. Accurate 3D registration of CT, PET and/or MR images of the brain. *J. Computer Assisted Tomography,* 13(1):20–26, 1989.

36. N. Ayache, J. D. Boissonnat, L. Cohen, B. Geiger, J. Levy-Vehel, O. Monga, and P. Sander. Steps toward the automatic interpretation of 3d images. In K.H. Hohne, editor, *NATO ARW, Vol F60, 3D Imaging in Medicine,* pages 107–120, Berlin, June 1990. Springer-Verlag.

37. G. M. Radack and N. I. Badler. Local matching of surfaces using a boundary-centered radial decomposition. *Computer Graphics, and Image Processing,* 45:380–396, 1989.

38. A. Gamboa-Aldeco, L. Fellingham, and G. Chen. Correlation of 3d surfaces from multiple modalities in medical imaging. In *SPIE Vol. 626, Medecine XIV/PACS IV,* pages 467–473, 1986.

39. S. Lavallee. *Geste Medico-Chirurgicaux Assistes par Ordinateur : Application a la Neurochirurgie Stereotaxique.* PhD thesis, Grenoble University, France, December 1989.

40. Y.L. Kergosien. Projection of smooth surfaces: stable primitives. In *MARI-COGNITIVA,* pages 447–454, Paris, France, 1987.

41. J. Barnes and P. Hut. A hierarchical $o(n \log n)$ force-calculation algorithm. *Nature,* 324:446–449, 4 December 1986.

Acknowledgements

We are very grateful to Professor Philippe Cinquin for his contribution to the idea of setting the 3-D/2-D matching problem in 3-D.
This research is financially supported by Digital Equipment Corporation and Sciences & Médecine-Safir.

A New Physically Based Model for Efficient Tracking and Analysis of Deformations

Chahab Nastar[1] and Nicholas Ayache[2]

[1] INRIA, Domaine de Voluceau, Rocquencourt
B.P. 105, 78153 Le Chesnay CEDEX, France.
[2] INRIA, B.P. 93, F-06902 Sophia-Antipolis, France.

Abstract. We present a physically-based deformable model which can be used to track and to analyze non-rigid motion of dynamic structures in time sequences of 2D or 3D medical images.

The model considers an object undergoing an elastic deformation as a set of masses linked by springs, where the classical natural lengths of the springs is set equal to zero, and is replaced by a set of constant equilibrium forces, which characterize the shape of the elastic structure in the absence of external forces.

This model has the extremely nice property of yielding dynamic equations which are linear and decoupled for each coordinate, whatever the amplitude of the deformation.

Compared to the former work of Terzopoulos and his colleagues [13, 27, 26, 16] and Pentland and his colleagues [22, 21, 23, 11], our model can be viewed as a continuation and unification ; it provides a reduced algorithmic complexity, and a sound framework for modal analysis, which allows a compact representation of a general deformation by a reduced number of parameters.

The power of the approach to segment, track and analyze 2-D and 3-D images is demonstrated by a set of experimental results on various complex medical images.

1 Introduction

We present a physically-based multidimensional deformable model, which can be used to track and to analyze non-rigid motion of dynamic structures in time sequences of 2D or 3D medical images.

In this model, an object is thought as a mesh of masses connected to eachother by springs. We show how the model is built, and how to integrate the dynamic equations through time, with the possibility to take into account some tracking contraints, as the knowledge of sparse anatomical features. We then dicuss the use of two different types of modes to describe the deformation, and propose to use the one which best fits our physical model. Finally, we illustrate by a set of experiments on synthetic and real data, the validity and the power of our approach to track and analyze the motion of anatomical structures both in two and three dimensions.

Concerning the tracking and the analysis of deformable objects, our work differs from but is related to the one of [17], who constrains locally the deformations to be conformal, and also to the one of [8, 1, 5], where tracking takes into account local differential properties of the surfaces. The principal warps analysis of [4] as well as the Fourier decomposition of [24] both show a similar spirit as our modal analysis. Finally the work of [6] and [9] provides an alternative way to segment volumetric images, but without explicitly trying to model the deformations.

2   The model

We consider both the surface and volumetric properties of the objects at hand. We restrict ourselves to elastic deformations, i.e. we assume the object recovers its reference configuration as soon as all applied forces causing deformation are removed. Modelling an elastic boundary M can be achieved by a mesh of n virtual masses on the contour, each mass being attached to its neighbors by springs of stiffness k and natural length l_0, as shown in figures 1 and 2. Similar discrete mass-spring models have been used in [26, 25, 29, 12]. These springs model the elastic *surface* properties of the object.

We can improve the modelling by attaching extra springs between non-neighbor nodes in order to model some *volumetric* elastic properties inside the object (see appendix A). These springs constrain the general form of the object within its deformation. The boundary M modelled as above will also be called

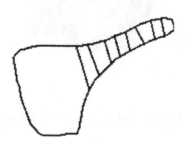

Fig. 1. A 2D valve model with surface and volume springs

structure. Such a structure can be easily deformed to match the contour of an object of interest, thus performing a *segmentation* step. Now if we take a set of images displaying the deformation of the object, structure M can also achieve simultaneously both *segmentation* and *tracking* of the object's surface through time (for more details, see [19]).

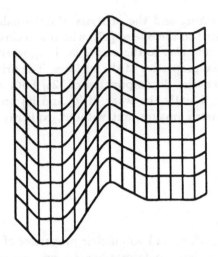

Fig. 2. A 3D mass-spring mesh with surface springs

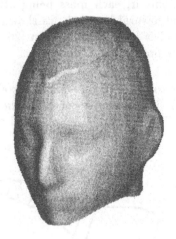

Fig. 3. Segmentation of a human head from a 3D magnetic resonance image

3 The governing equation

The system under study is composed of n masses that are positionned at time t on the points $((M_1)_t, (M_2)_t, \ldots, (M_n)_t)$. Let :

$$M_t = [(M_1)_t, (M_2)_t, \ldots, (M_n)_t]^T$$

The evolution of the structure is governed by the fundamental equation of dynamics :

$$F_i = m_i a_i \qquad i = 1, \ldots, n$$

where m_i is the mass of point M_i and a_i its acceleration under total load F_i.

Now, what are the applied forces to the point M_i at time t? First, there is the elastic force due to its neighbors :

$$F_e(M_i, t) = -k \sum_{j \in C_i} U_{ij} - l_0 \sum_{j \in C_i} \frac{U_{ij}}{\|U_{ij}\|} \tag{1}$$

where l_0 is the natural length of the springs, $U_{ij} = (M_j M_i)_t$ is the vector separation of nodes i and j at time t, and C_i is the set of nodes connected to node i.

Then, a damping force can be considered. This force is generally set proportional to the point velocity v_i :

$$F_d(M_i, t) = -c_i v_i$$

Moreover, an external load $F_{ext}(M_i, t)$ acts on each node M_i. Note that the structure must be given an initial position at time t_0. Since we want the structure to hold in this position, we have to apply on each node a force F_{eq} so that time t_0 is an equilibrium state of the structure. This force is similar to the force our fingers apply to an elastic in order to give a certain form to it. Thus, at initial time, we have :

$$F_{eq}(M_i) + F_e(M_i, t_0) = 0 \tag{2}$$

that is, the sum of the forces acting on each is zero. We assume that at any future time this equilibrium force is constant.

Hence, the total load at time t is for each node M_i :

$$F(M_i, t) = F_e(M_i, t) + F_d(M_i, t) + F_{ext}(M_i, t) + F_{eq}(M_i)$$

The governing equation of the structure is :

$$F_e(M_i, t) + F_d(M_i, t) + F_{ext}(M_i, t) + F_{eq}(M_i) = m_i a_i$$

This equation, expressed for all n nodes, leads to a *nonlinear* system of *coupled* differential equations (for each node, the x, y and z displacements are coupled, and the displacement of a node depends on its neighbors' displacement, as it is clearly shown in equation (1)). One possible approach is the solving of these complex nonlinear coupled equations by an iterative procedure [26].

We propose in this paper to set $l_0 = 0$, an assumption which does not restrict the structure's arbitrary initial configuration (allowed by the constant equilibrium force field F_{eq}). The advantage of this assumption is that it considers the model in the linear elasticity framework. As a consequence, we end up with a set of *linear* differential equations with node displacements *decoupled* in each coordinate, whatever the magnitude of the displacements. Moreover, these linear equations are a prerequisite to further quantitative analysis of the motion (see chapter 7). On the other hand, our approximation is valid only if the springs orientations undergo small angular variations (less than 15 degrees), so that our assumption of constant F_{eq} holds. Similar limitations can be found in the model described in [23]. A direction of research to avoid this may be the use of the model described in [16].

More precisely, setting $l_0 = 0$ and denoting $U_i(t) = (M_i)_t - (M_i)_{t_0}$ we can express the elastic force in terms of nodal displacements :

$$F_e(M_i, t) = -k \sum_{j \in C_i} U_{ij}$$

$$F_e(M_i, t) = F_e(M_i, t_0) - k[card(C_i)\, U_i(t) - \sum_{j \in C_i} U_j(t)]$$

where $card(C_i)$ is the cardinality of C_i (i.e. the number of nodes connected to node i). Remembering equation (2), the governing equation becomes :

$$-k[card(C_i)\, U_i(t) - \sum_{j \in C_i} U_j(t)] - c_i \dot{U}_i(t) + F_{ext}(M_i, t) = m_i \ddot{U}_i(t)$$

These equations can now be written in a matrix form :

$$M\ddot{U} + C\dot{U} + KU = F_t \qquad (3)$$

where K is the stiffness matrix of the element (see appendix A), C its damping matrix, M its mass matrix, and F_t is the external force field at time t, and U is the nodal displacements vector ($U = M_t - M_{t_0}$). Note that in the above formulation the stiffness matrix is constant (i.e. the same at each time step), and no recomputation is necessary. As to F_{eq}, its computation is straightforward as soon as the initial shape is chosen (see equation (2)) ; however F_{eq} does not appear in the governing equation : this force field can now be viewed as an *internal* force field which does not need to be computed.

In 3D, the above matrix-form equation represents $3n$ coupled second-order ordinary differential equations. However, by setting $l_0 = 0$, this equation is separable in three n order equations with respect to x, y and z. Therefore, from now on, except when specified, vectors and matrices will be of order n. For instance, U will refer indifferently to the x, y or z nodal displacements vector.

4 An elastic potential

4.1 General form of the potential

We make the common assumption that the external force field at time t, derives from a potential field V_t :

$$F_t = -\nabla V_t \qquad (4)$$

Various potential fields can successfully solve the problem. In the "snake" formulation [13, 26], in order to have the snake attracted to contours with large image gradients, the potential on any point M_i is set to :

$$V(M_i) = -\|\nabla G_\sigma \star I(M_i)\|$$

where $G_\sigma \star$ denotes convolution with a gaussian smoothing filter of width σ, so that the edges of the image can attract the snake from a distance. Note that the farther the snake from the edges, the weaker the attraction force field.

In our formulation, we need a potential field that can *a posteriori* describe which nodal displacements are the most likely. One interesting choice for the potential is :

$$V(M_i) = \frac{1}{2}k_p\|P_iM_i\|^2 \tag{5}$$

where P_i is the closest boundary point to M_i. The algorithm for obtaining this potential on each point of an image is described in [7] (see also [14, 15]).

Another advantage of this potential is that it can be represented by a spring of natural length zero and of stiffness k_p joining M_i to P_i :

$$F_p(M_i) = -\nabla V(M_i) = k_p M_i P_i$$

Hence, the whole model remains elastic. Moreover, the farther the structure from the edges, the larger the attraction force. This procedure speeds up the segmentation of the image edges by the deformable structure.

Thus, we have defined n springs of natural length zero and of stiffness k_p joining structure nodal points M_i to their closest boundary point P_i. The external force field at time t is :

$$F_t = k_p D_t$$

where $D_t = [(M_1P_1)_t, (M_2P_2)_t, \ldots, (M_nP_n)_t]^T$ is a distance field which may not be unique (several structure points may have the same closest boundary point).

The governing equation is, for this choice of the potential field :

$$M\ddot{U} + C\dot{U} + KU = k_p D_t \tag{6}$$

4.2 Automatic selective potential

Suppose that the displacement of node i_0 is known (M_{i_0} matches M'_{i_0}) by an alternative method (for example by use of anatomical or artificial landmarks such as high curvature points, see [1, 5]). The displacement of landmark M_{i_0} has to be computed separately, in other terms, M_{i_0} must be submitted to another potential field that the non-landmark nodes, having it attracted by M'_{i_0}.

This can be achieved automatically with 2D and 3D feature extraction algorithms [18, 28]. We set the potential differently on every point $(M_i)_t$, depending on $(M_i)_t$ being a landmark or not. If $(M_{i_0})_t$ is a landmark, we find its new deformed position, namely, the point M'_{i_0} that has the same feature (for instance, $(M_{i_0})_t$ and M'_{i_0} are both maximum curvature points on the surface). We submit M_{i_0} to the potential field :

$$V(M_{i_0})_t = \frac{1}{2}\alpha\|M'_{i_0}(M_{i_0})_t\|^2$$

For the non-landmark nodes $(M_i)_t$, the potential is computed as :

$$V(M_i)_t = \frac{1}{2}k_p\|(P_i)_t(M_i)_t\|^2$$

where $(P_i)_t$ is the closest boundary point to $(M_i)_t$ at time t, and $\alpha \gg k_p$. Intuitively, the physical interpretation of the landmark constraint for M_{i_0} can be seen as attaching a stiff spring between M_{i_0} and M'_{i_0}, and discarding the spring between M_{i_0} and P_{i_0} (figure 4).

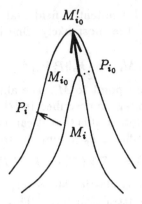

Fig. 4. Attraction force for landmark node M_{i_0} and for non-landmark node M_i

5   Direct integration of the governing equation

As the distance field may not be unique, there may be several acceptable solutions to the tracking problem. This implies that we have to give the structure a *reasonable* position at initial time $t_0 = 0$, that is, the initialization has to minimize the risk of conflicts.

The governing equation can then be directly integrated through time by various methods [3, 26]. One of the simplest and quickest ones is the explicit Euler method :

$$\begin{cases} \ddot{U}_t = M^{-1}(F_t - C\dot{U}_t - KU_t) \\ \dot{U}_{t+\Delta t} = \dot{U}_t + \Delta t \ddot{U}_t \\ U_{t+\Delta t} = U_t + \Delta t \dot{U}_{t+\Delta t} \end{cases}$$

where the damping, inertial and stiffness constants are chosen so that the system is overdamped.

6   Results on 2D objects

6.1   Segmentation

Figure 5 outlines the importance of the volume springs. A classical curvilinear snake cannot segment the simulated valve given this initial position : it collapses against the bottom boundary of the second valve contour (figure 5.a and 5.b). Volume springs prevent the elastic structure from collapsing (figure 5.c and 5.d). A hundred of nodal points were used for the modelling.

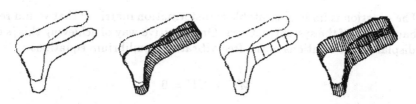

Fig. 5. Segmentation of the mitral valve without (a,b) and with (c,d) volume springs

6.2 Tracking

Figure 6 shows the modification of the nodal displacements when the displacement of one of the nodes is known (compare the modification of the displacement field betwwen figures 6.b and 6.d)). This result is an alternative to the results obtained with another energy-minimizing procedure [5]. Sixty nodal points were used for the modelling.

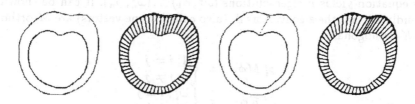

Fig. 6. Unconstrained (a,b) and constrained (c,d) tracking of a 2D contour

7 Modal analysis

This well-known approach in the field of mechanical engineering [3], was brought to the field of computer vision by Pentland's team [23, 11]. We outline the general principle and present our formulation, which differs from the one of Pentland in being better adapted to the underlying physical model we are using.

7.1 Change of basis

Instead of solving directly the equilibrium equation (3), one can transform it by a change of basis :

$$U = P\tilde{U}$$

where P is the square n order nonsingular transformation matrix to be determined, and $\tilde{U}$ is referred to as the *generalized displacements* vector (see [3]).

The question is finding a suitable transformation matrix P that would reduce the bandwidth of the system matrices. One effective way of achieving this is using the displacement solutions of the free vibration equilibrium equations :

$$M\ddot{U} + KU = 0 \tag{7}$$

where M is the mass matrix of the structure. The solution to (7) can be postulated to be of the form :

$$U = \phi \sin \omega(t - t_0)$$

where ϕ is a vector of order n, t the time variable, t_0 a time constant, and ω the frequency of vibration of the vector ϕ. We can now substitute this expression of U in (7) to determine ϕ and ω, which leads to the generalized eigenproblem :

$$K\phi = \omega^2 M\phi \tag{8}$$

This equation yields n eigensolutions $(\omega_1^2, \phi_1) \ldots (\omega_n^2, \phi_n)$. It can be shown (see appendix B) that the eigenvectors ϕ_i (also called shape vectors) are M-orthonormal and K-orthogonal :

$$
\begin{aligned}
\phi_i^T M \phi_j &= \begin{cases} 1; i = j \\ 0; i \neq j \end{cases} \\
\phi_i^T K \phi_j &= \begin{cases} \omega_i^2; i = j \\ 0; \ i \neq j \end{cases} \\
0 \leq \omega_1^2 &\leq \omega_2^2 \leq \ldots \leq \omega_n^2
\end{aligned}
$$

The vectors $(\phi_i)_{i=1,\ldots,n}$ can then be referred to as the structure's *eigenbasis*. The structure's eigenbasis is determined as soon as the mass matrix and the stiffness matrix of the structure are defined.

The former equations can be rewritten in a matrix form :

$$\tilde{M} = \Phi^T M \Phi = I_n \tag{9}$$
$$\tilde{K} = \Phi^T K \Phi = \Omega^2 \tag{10}$$

where Φ is the matrix whose columns are the eigenvectors ($\Phi = [\phi_1, \phi_2, \ldots, \phi_n]$) and Ω^2 is a diagonal matrix which stores the eigenvalues ω_i^2 on its diagonal ($\Omega^2 = diag(\omega_1^2, \omega_2^2, \ldots, \omega_n^2)$). The matrix Φ is then chosen to be the transformation matrix P :

$$U = \Phi \tilde{U} = \sum_{i=1}^{n} \tilde{u}_i \phi_i \tag{11}$$

Equation (11) is referred to as the modal superposition equation. $\tilde{u}_i$ is the ith-mode *amplitude* within displacement U.

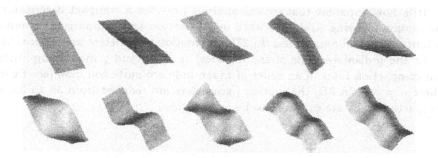

Fig. 7. Energy-increasing eigenmodes of a flat element

7.2 Decoupling the equations of motion

Premultiplying the governing equation (3) by Φ^T and expressing the displacement vector in terms of the generalized displacement vector leads to :

$$\ddot{\tilde{U}} + \tilde{C}\dot{\tilde{U}} + \Omega^2\tilde{U} = \Phi^T F \qquad (12)$$

Hence, if the matrix $\tilde{C} = \Phi^T C \Phi$ is diagonal, then the above matrix-form equations decouple into n scalar equations :

$$\ddot{\tilde{u}}_{t,i} + \tilde{c}_i\dot{\tilde{u}}_{t,i} + \omega_i^2\tilde{u}_{t,i} = \tilde{F}_{t,i} \qquad i = 1,\ldots,n. \qquad (13)$$

Solving these equations at time t leads to $(\tilde{u}_{t,i})_{i=1,\ldots,n}$, and the displacement U_t of the structure's nodes is obtained by the modal superposition equation.

7.3 Modal approximation

In practice, we wish to approximate the nodal points displacements U by picking up p significant modes, where $p \ll n$:

$$U_t \approx \sum_{i=1}^{p} \tilde{u}_{t,\sigma(i)}\phi_{\sigma(i)} \qquad (14)$$

where σ is a suitably chosen permutation.

The $\left(\phi_{\sigma(i)}\right)_{i=1,\ldots,p}$ are the *reduced eigenbasis* of the structure.

This is the major advantage of modal analysis : it provides an approximate but quite accurate solution by selecting a few number of modes. The participation of each mode to the motion is ordered. In those terms, we can compare modal analysis to principal components analysis. Whereas in principal components analysis the object, which is in a high order space, is displayed after being projected into a lower order subspace maximizing its spreading, modal analysis projects the nodal displacements vector U (which is in a n order space) into a subspace of much lower order p, with minimum loss of accuracy.

It is now apparent that modal analysis provides a compact description of the motion allowing straightforward interpretation and comparison of motions in terms of modal amplitudes. In 2D, the motion parameters are reduced from $2n$ (x and y displacements of the n nodes) to $2p$ (x and y modal amplitudes), and compaction rates of an order of magnitude are quite common (see figure 8 where $n = 80$). In 3D, the motion parameters are reduced from $3n$ to $3p$, and compaction rates are expected to be even higher.

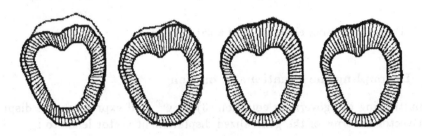

Fig. 8. Superposition of high amplitude modes. a. 2 modes (compaction : 80, recovery : 50.6 %) b. 4 modes (compaction : 40, recovery : 68.7 %) c. 20 modes (compaction : 8, recovery : 98.2 %) d. All 160 modes (compaction : 1, recovery : 100.0 %)

7.4 Selecting the most significant modes

In this section we concentrate on the choice of the p modes that can most accurately describe the object's motion.

The most accurate choice is the choice of the p modes of *highest amplitude*, having computed $\tilde{u}_{t,i}$ by equation (13). Instead of choosing the number p of significant modes, we can rather choose a constant μ ($0 < \mu \leq 1$) which represents the confidence we need for the approximation of the displacement :

$$\frac{\sum_{i=1}^{p} |\tilde{u}_{t,\sigma(i)}|}{\sum_{i=1}^{n} |\tilde{u}_{t,i}|} \geq \mu \tag{15}$$

Instead of sorting the modal amplitudes, and remembering equation (13), we can choose the p *lowest frequency modes* to approximate the displacement ($\sigma = Id$, see appendix C). Other advantages of this method are : first, whatever the motion, the reduced eigenbasis is constant for a given structure ; then, only $3p$ equations instead of $3n$ of the form of equation 13 are to be solved.

7.5 Qualitative modes (Pentland's approach)

In this section vectors and matrices are of order $3n$.
One can choose another suitable matrix P to perform the change of basis :

$$U = P\hat{U}$$

A modal superposition can be performed, if we set $P = [\psi_1, \ldots, \psi_{3n}]$ where it is assumed that each mode ψ_i is orthogonal to all others, but is not necessarily an eigenmode of the element :

$$U = \sum_{i=1}^{3n} \hat{u}_i \psi_i$$

The modes ψ_i can be set to be elementary displacements referred to as : translation, rotation, scaling, shearing... (see appendix D). They are "qualitative modes".

In general, q ($q < n$) qualitative modes are considered. Denoting $\Psi = [\psi_1, \ldots, \psi_q]$ the change of basis leads to the equation

$$\Psi^T M \Psi \ddot{\hat{U}} + \Psi^T C \Psi \dot{\hat{U}} + \Psi^T K \Psi \hat{U} = \Psi^T F \tag{16}$$

But the change of basis is interesting when matrices $\Psi^T K \Psi, \Psi^T M \Psi$ and $\Psi^T C \Psi$ are diagonal, so that the q above equations decouple. Pentland's team proposes to set directly these matrices to some diagonal matrices [23, 11], since the stiffness matrix representing the nodes connections has not been calculated.

As we have calculated the stiffness matrix, we prefer to make use of the calculated node displacements vector and project it in the q-dimensional space defined by the q qualitative modes :

$$proj(U) = \sum_{i=1}^{q} \hat{u}_i \psi_i$$

The amplitudes $\hat{u}_i$ can then be easily determined, since the orthonormality of qualitative modes (with respect to scalar product) can generally be ensured [20] :

$$\hat{u}_i = proj(U).\psi_i$$

As the qualitative modes often represent global geometrical transforms, they can be assumed to be of low natural frequency, and the approximation of U by $proj(U)$ is satisfactory, mainly if the displacement norm $\|U\|$ is small.

7.6 Eigenmodes versus qualitative modes

Eigenmodes form a complete basis in which the motion can be expressed accurately or approximately. Their *physical* interpretation is clear : they are the monofrequency vibration harmonics of the elastic structure. Moreover, our model provides decoupled eigenmodes with respect to x, y and z, reducing the dimensions of the handled matrices from $3n \times 3n$ to $n \times n$. The eigenmodes seem to be the best modal basis for computer vision.

On the other hand, qualitative modes do not share the physical interpretation of the eigenmodes. They consist mainly in a *geometric* decomposition of non-rigid motion, but which is not related to the model. They are derived from our common vocabulary, and can be quite easily computed via a polynomial approximation [23]. They seem to be more dedicated to computer graphics, where certain types of geometric deformations are to be synthetized.

In figures 9 and 10, the motion of the valve with volume springs is analyzed in modal space. The eigenmodes and their superposition, as well as the qualitative modes and their superposition are displayed. Note the error in the recovery of the motion using the qualitative modes (figure 10) versus the accuracy of the superposition eigenmodes using the same number of modes (figure 9). This comes from the fact that our eigenmodes better correspond to the physical model we have developped for the motion. On the other hand, qualitative modes are expressed in a reference frame intrinsic to the object, a nice property that our own eigenmodes can share provided that we choose the intrinsic reference frame of the model (using the center of inertia and the axes of inertia).

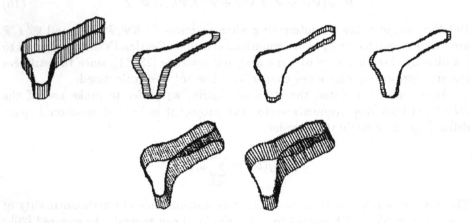

Fig. 9. Four high amplitude eigenmodes of the valve (a,b,c,d). Their superposition (e) and the expected result (f)

7.7 Applications

One of the applications of modal approximation is *motion correction*. In order to maintain an object's geometry within its displacement, we can approximate the nodal displacements vector by superimposing a few number of low-frequency modes, then substitute the approximated displacement to the original one. The displacement field is smoothed and the deformed object has global geometric properties similar to the ones of the undeformed object (figure 11).

The reduced eigenbasis is then a good low order basis enabling straightforward *comparisons* of different types of motions. Moreover, the representation of

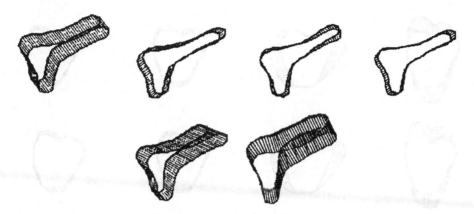

Fig. 10. Four qualitative modes of the valve (a,b,c,d). Their superposition (e) and the expected result (f)

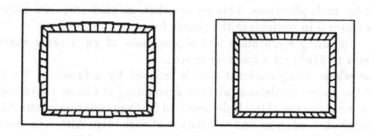

Fig. 11. Motion correction of a rectangle using low frequency mode shapes

U in modal space is stable when the number n of structure points varies. In other terms, the modal representation is not too sensitive to the sampling of the boundary (which is always arbitrary), which is a very important property (figure 12).

Note that the mode shapes depend upon the original model ; in other words, had we chosen another way of connecting the nodal points (another K matrix), the eigenmodes would have been different. Note also that the elastic model we are using has to be linearly elastic so that the eigenmodes are constant vectors. For a nonlinear elastic model, the stiffness matrix has to be recomputed at each step, and the eigenmodes become time varying : modal analysis is then impractical.

8 Computational cost

We consider the general case where matrix K is only symmetric. The choice of $l_0 = 0$ which enables decoupled equations in x, y and z reduces the matrix vector multiplication complexity from $O(9n^2)$ for $3n$-order matrices to $O(3n^2)$

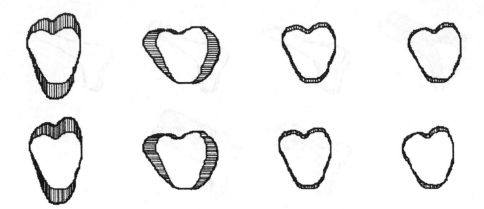

Fig. 12. Four high amplitude eigenmodes with 80 and 95 nodal points

for 3 n-order multiplications. This means that, at each step, we reduce CPU time by a factor 3 in computing the elastic force.

Then, computing eigenvalues and eigenvectors of an n order matrix is in $O(n^3)$ versus $O(27n^3)$ for a $3n$-order matrix.

On the whole, computational cost is reduced by a factor of $3 \times 27 = 81$ compared the same calculations without decoupling the three coordinates.

Finally, we make use of the Euler method for the resolution of our differential equations, which needs only one calculation at each step. Although the stability of this method is limited, it is less costly than a more stable method like Runge-Kutta, which can be used if necessary.

9 Experimental results

9.1 3D segmentation of human head

A mass spring mesh is used to segment a 3D magnetic resonance image of a head. The resolution of the 3D image is $158 \times 158 \times 158$.

Figures 13, 14 and 15 show the segmentation of the human head by a deformable cylindric mesh of $159 \times 70 = 11130$ nodes. The initial mesh is given by the user as a 2D curve, that will be repeated on the 70 plans to form a cylindric mesh. The solution is displayed after 50 iterations, each iteration being performed in a few seconds CPU on a workstation, although no optimization of the code is done yet.

9.2 2D tracking of human left ventricle

We have tested our method on a set of ultrasound images of the human heart's left ventricle. The tracking of the mitral valve is indeed a problem of major interest in medical imaging. First, a polar edge extraction is performed on the

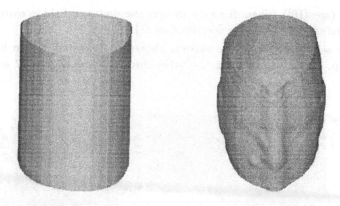

Fig. 13. Initial mesh and solution after 5 iterations

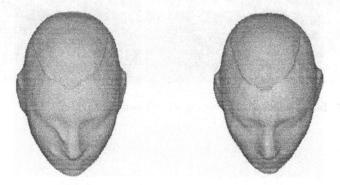

Fig. 14. Solution after 20 and 50 iterations

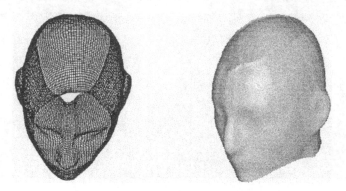

Fig. 15. Segmentation of the human head (11130 nodes)

255

images (see [10]). Then, for each image, the distance field is computed on every pixel, using the algorithm described in [7].

The segmentation of the valve is shown in figure 16. It can be performed in one step. Then we can track the valve through time (figure 17 and 18).

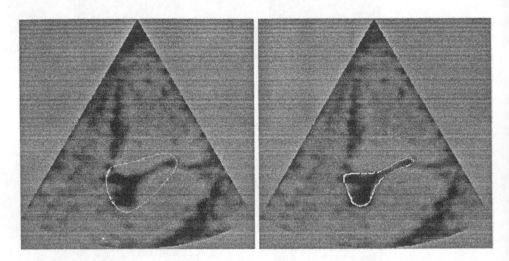

Fig. 16. Initial segmentation of the valve's contour

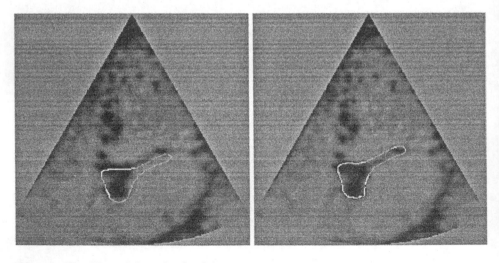

Fig. 17. Tracking of the mitral valve

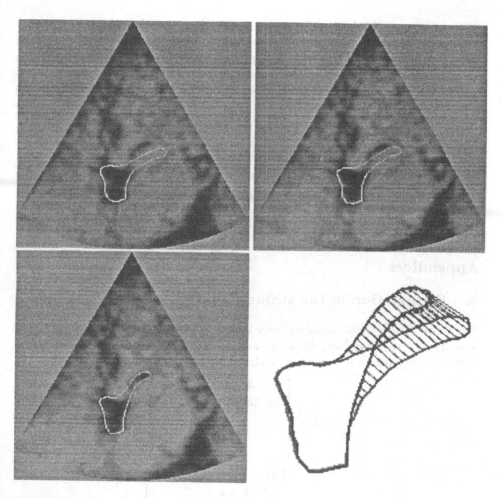

Fig. 18. Tracking of the mitral valve

10 Conclusion

We have presented an elastic model enabling fast segmentation and tracking of 2D and 3D images. The high flexibility of the method makes it easy to use for modelling various types of elastic objects. Non-rigid motion is modelled continuously thanks to the use of physical entities (mass, velocity, acceleration) and the trajectory of each node can be displayed.

The stiffness matrix remains constant at all iterations, so that, for a given structure, the modal system is set up once for many fits to different data sets. The computations are decoupled in x,y and z, and The external potential field is selective, depending on the node being a landmark or not. Thanks to the introduction of equilibrium forces F_{eq}, it is possible to set the natural lengths of the springs to zero ; this yields a set of *linear* differential equations *decoupled* in

x, y and z, even for large deformations. Therefore the algorithmic complexity is reduced by almost two orders of magnitude. Also, we showed how to take into account additional tracking constraints between irregular features. Finally, we adopted modal analysis of the motion to better fit the underlying physical model, and showed its power for a compact description and/or smoothing of a complex deformation.

Acknowledgements

We wish to thank Gregoire Malandain for stimulating discussions about the subject and for his implementation of the force field, and Siemens France for providing the 3D MRI data. This work was supported in part by a grant from Digital Equipment Corporation.

Appendices

A Formulation of the stiffness matrix

For a two-dimensional structure with identical springs of stiffness k joining a mass to its immediate neighbors, the monodimensional n order stiffness matrix is respectively for an open and for a closed structure :

$$K_{open} = \begin{bmatrix} k & -k & & & \\ -k & 2k & -k & & \\ & & & -k & 2k & -k \\ & & & & -k & k \end{bmatrix} \tag{17}$$

$$K_{closed} = \begin{bmatrix} 2k & -k & & & -k \\ -k & 2k & -k & & \\ & & & -k & 2k & -k \\ -k & & & & -k & 2k \end{bmatrix} \tag{18}$$

In the 3D case, the stiffness matrix contains submatrices that have the form described above.

We can model the structure in a more complex way, for instance by attaching extra springs of stiffness k_{extra} between non-neighbor nodes i and j. The stiffness matrix is then modified as follows :

$$(K_{ii})_{extra} = K_{ii} + k_{extra}$$
$$(K_{jj})_{extra} = K_{jj} + k_{extra}$$
$$(K_{ij})_{extra} = K_{ij} - k_{extra}$$
$$(K_{ji})_{extra} = K_{ji} - k_{extra}$$

Note that the stiffness matrices are non-definite positive ($KT = 0$, with $T = [1, 1, \ldots, 1]^T$).

B Orthogonality of mode shapes

An important property of mode shapes is orthogonality, which is discussed here. For the ith and jth natural frequencies ω_i and ω_j and the ith and jth mode shapes ϕ_i and ϕ_j, we have :

$$K\phi_i = \omega_i^2 M\phi_i \tag{19}$$

$$K\phi_j = \omega_j^2 M\phi_j \tag{20}$$

Premultiplying equation (19) by ϕ_j^T and equation (20) by ϕ_i^T leads to :

$$\phi_j^T K\phi_i = \omega_i^2 \phi_j^T M\phi_i \tag{21}$$

$$\phi_i^T K\phi_j = \omega_j^2 \phi_i^T M\phi_j \tag{22}$$

Taking the transpose of equation (22), and remembering that mass and stiffness matrices are symmetric, we obtain :

$$\phi_j^T K\phi_i = \omega_j^2 \phi_j^T M\phi_i \tag{23}$$

Comparing now equation (21) and equation (23) yields :

$$(\omega_i^2 - \omega_j^2)\phi_j^T M\phi_i = 0 \tag{24}$$

Since $\omega_i^2 \neq \omega_j^2$, we have :

$$\phi_j^T M\phi_i = 0 \qquad \forall i \neq j$$

M being a positive definite matrix, we can norm the mode shapes to obtain :

$$\phi_i^T M\phi_i = 1 \qquad \forall i$$

The mode shapes are then M-orthonormal.

Using M-orthonormality in equation (21) leads to the K-orthogonality of mode shapes :

$$\phi_j^T K\phi_i = 0; \qquad \forall i \neq j$$
$$\phi_i^T K\phi_i = \omega_i^2; \qquad \forall i$$

C Selecting the low natural frequencies

Using an easy example, we discuss here the choice of mode shapes that have the lowest natural frequencies for the approximation of the displacement vector. For the detailed equations, see [3].

Consider a clamped beam. Figure 19 shows the aspect of a low order and a high order mode.

C.1 Frequential concordance

Consider a concentrated load in the middle of the beam. If the load varies slowly, the natural high frequencies are not amplified. This is illustrated in figure 20.

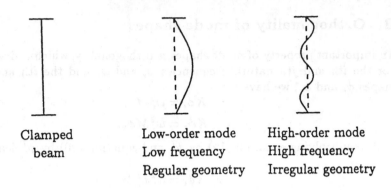

| Clamped
beam | Low-order mode
Low frequency
Regular geometry | High-order mode
High frequency
Irregular geometry |

Fig. 19. a clamped beam

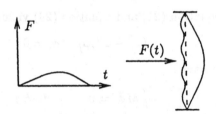

Slow load : Low frequency modes excited

Pseudo-static effect on high frequency modes

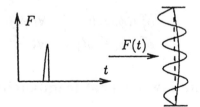

Fast load : High frequency modes excited

No exciting effect on Low frequency modes

Fig. 20. Frequential concordance

C.2 Geometric participation

If the load is equally distributed along the beam, the natural high frequencies, whose geometry is irregular, are not excited. The amount of work for force F (that is $\Phi^T F$ for displacement Φ) is negligible for these frequencies (figure 21).

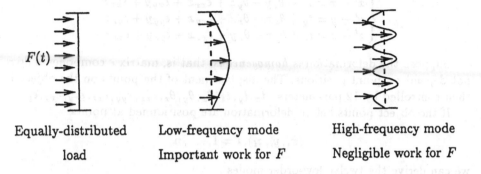

| Equally-distributed | Low-frequency mode | High-frequency mode |
| load | Important work for F | Negligible work for F |

Fig. 21. Geometric participation

D Qualitative modes

As described before, the qualitative modes ψ_i of an object are generally not the solution of the eigenproblem $K\phi = \omega^2 M\phi$.

Their main advantage is that they can be described by common terms, such as translation, rotation (rigid-body modes), scaling, shearing (first order modes), tapering, pinching, bending (second-order modes)...

To intoduce the qualitative modes, consider the general non-rigid motion of an object. It can be represented by :

$$\begin{cases} M(x,y,z) \rightarrow M'(x',y',z') \\ M' = rd\,M + t \end{cases}$$

Where r is a rotation matrix of angles $\theta_x, \theta_y, \theta_z$, t is a translation vector ($t = t_x + t_y + t_z$), and d is a symmetric deformation matrix.

Suppose now that *the displacement is "small"*. Then rotation can be linearized ($|\theta| \ll \pi$) :

$$r = I + \rho = \begin{bmatrix} 1 & 0 & 0 \\ 0 & 1 & 0 \\ 0 & 0 & 1 \end{bmatrix} + \begin{bmatrix} 0 & -\theta_z & -\theta_y \\ \theta_z & 0 & -\theta_x \\ \theta_y & \theta_x & 0 \end{bmatrix}$$

And deformation can be decomposed into :

$$d = I + \epsilon$$

where ϵ is the symmetric strain matrix.

A first order approximation of the motion leads to :

$$\begin{cases} M(x, y, z) \rightarrow M'(x', y', z') \\ M' - M = (\rho + \epsilon)\, M + t \end{cases}$$

More precisely, in $3D$, we have :

$$\begin{cases} x' - x = t_x - \theta_z y - \theta_y z + \epsilon_{xx} x + \epsilon_{xy} y + \epsilon_{xz} z \\ y' - y = t_y + \theta_z x - \theta_x z + \epsilon_{xy} x + \epsilon_{yy} y + \epsilon_{yz} z \\ z' - z = t_z + \theta_y x + \theta_x y + \epsilon_{xz} x + \epsilon_{yz} y + \epsilon_{zz} z \end{cases}$$

Suppose the deformation is *homogeneous*, that is, matrix ϵ components does not depend on point positions. The displacement of the points on the object is then controlled by 12 parameters : $t_x, t_y, t_z, \theta_x, \theta_y, \theta_z, \epsilon_{xx}, \epsilon_{yy}, \epsilon_{zz}, \epsilon_{xy}, \epsilon_{xz}, \epsilon_{yz}$.

If the object points before deformation are positionned at points

$$(x_i, y_i, z_i), i = 1, \ldots, n$$

we can derive the twelve low-order modes :

$$\begin{aligned}
\psi_1^T &= [1, 0, 0, 1, 0, 0, \ldots, 1, 0, 0] & & x \text{ translation} \\
\psi_2^T &= [0, 1, 0, 0, 1, 0, \ldots, 0, 1, 0] & & y \text{ translation} \\
\psi_3^T &= [0, 0, 1, 0, 0, 1, \ldots, 0, 0, 1] & & z \text{ translation} \\
\psi_4^T &= [-y_1, x_1, 0, -y_2, x_2, 0, \ldots, -y_n, x_n, 0] & & z \text{ rotation} \\
\psi_5^T &= [-z_1, 0, x_1, -z_2, 0, x_2, \ldots, -z_n, 0, x_n] & & y \text{ rotation} \\
\psi_6^T &= [0, -z_1, y_1, 0, -z_2, y_2, \ldots, 0, -z_n, y_n] & & x \text{ rotation} \\
\psi_7^T &= [x_1, 0, 0, x_2, 0, 0, \ldots, x_n, 0, 0] & & x \text{ scaling} \\
\psi_8^T &= [0, y_1, 0, 0, y_2, 0, \ldots, 0, y_n, 0] & & y \text{ scaling} \\
\psi_9^T &= [0, 0, z_1, 0, 0, z_2, \ldots, 0, 0, z_n] & & z \text{ scaling} \\
\psi_{10}^T &= [y_1, x_1, 0, y_2, x_2, 0, \ldots, y_n, x_n, 0] & & z \text{ shearing} \\
\psi_{11}^T &= [z_1, 0, x_1, z_2, 0, x_2, \ldots, z_n, 0, x_n] & & y \text{ shearing} \\
\psi_{12}^T &= [0, z_1, y_1, 0, z_2, y_2, \ldots, 0, z_n, y_n] & & x \text{ shearing}
\end{aligned}$$

Thus, in 3D, assuming that the displacement is small and the strain matrix is homogeneous, only twelve modes are necessary for parameterizing the displacement. Note that these assumptions are restrictive, and more polynomial modes have generally to be considered [21], and their orthonormality must be ensured.

References

1. A. Amini, R. Owen, L. Staib, P. Anandan, and J. Duncan. *Non-rigid motion models for tracking the left ventricular wall.* Lecture notes in computer science: Information processing in medical images. Springer-Verlag, 1991.
2. N. Ayache, I. Cohen, and I. Herlin. Medical image tracking. In A. Blake and A. Yuille, editors, *Active Vision*, chapter 17. MIT-Press, 1992.
3. Klaus-Jurgen Bathe. *Finite Element Procedures in Engineering Analysis.* Prentice-Hall, 1982.

4. Fred L. Bookstein. Principal warps: Thin-plate splines and the decomposition of deformations. *IEEE Transactions on Pattern Analysis and Machine Intelligence*, PAMI-11(6):567–585, June 1989.

5. Isaac Cohen, Nicholas Ayache, and Patrick Sulger. Tracking points on deformable objects. In *Proceedings of the Second European Conference on Computer Vision (ECCV) 1992*, Santa Margherita Ligure, Italy, May 1992.

6. Isaac Cohen, Laurent D. Cohen, and Nicholas Ayache. Using deformable surfaces to segment 3-D images and infer differential structures. *Computer Vision, Graphics, and Image Processing*, 1992.

7. P. E. Danielsson. Euclidean distance mapping. *Computer Vision, Graphics, and Image Processing*, 14:227–248, 1980.

8. J.S. Duncan, R.L. Owen, L.H. Staib, and P. Anandan. Measurement of non-rigid motion using contour shape descriptors. In *Proc. Computer Vision and Pattern Recognition*, pages 318–324. IEEE Computer Society Conference, June 1991. Lahaina, Maui, Hawaii.

9. Andre Gueziec. Large deformable splines : Crest lines and matching. In *Proceedings of the Fourth International Conference on Computer Vision*, Berlin, May 1993.

10. I.L. Herlin and N. Ayache. Features extraction and analysis methods for sequences of ultrasound images. In *Proceedings of the Second European Conference on Computer Vision (ECCV) 1992*, Santa Margherita Ligure, Italy, May 1992.

11. Bradley Horowitz and Alex Pentland. Recovery of non-rigid motion and structure. In *Proc. Computer Vision and Pattern Recognition*, pages 325–330. IEEE Computer Society Conference, June 1991. Lahaina, Maui, Hawaii.

12. Wen-Chen Huang and Dmitry B. Goldgof. Adaptive-size physically based models for nonrigid motion analysis. In *Proc. Computer Vision and Pattern Recognition*, pages 833–835. IEEE Computer Society Conference, June 1992. Champaign, Illinois.

13. Michael Kass, Andrew Witkin, and Demetri Terzopoulos. Snakes: Active contour models. *International Journal of Computer Vision*, 1:321–331, 1987.

14. F. Leitner, I. Marque, S. Lavallée, and P. Cinquin. Dynamic segmentation: finding the edge with snake-splines. In *Proceedings of International Conference on Curves and Surfaces*, pages 1–4, Chamonix, France, June 1990. Academic Press.

15. F. Leymarie and M.D. Levine. Simulating the grassfire transform using an active contour model. *IEEE Transactions on Pattern Analysis and Machine Intelligence*, PAMI-14, January 1992.

16. Dimitri Metaxas and Demetri Terzopoulos. Constrained deformable superquadrics and nonrigid motion tracking. In *Proc. Computer Vision and Pattern Recognition*, pages 337–343. IEEE Computer Society Conference, June 1991. Lahaina, Maui, Hawaii.

17. Sanjoy K. Mishra, Dmitry B. Goldgof, and Thomas S. Huang. Motion analysis and epicardial deformation estimation from angiography data. In *Proc. Computer Vision and Pattern Recognition*, pages 331–336. IEEE Computer Society Conference, June 1991. Lahaina, Maui, Hawaii.

18. Olivier Monga, Serge Benayoun, and Olivier D. Faugeras. From partial derivatives of 3d density images to ridge lines. In *Proc. Computer Vision and Pattern Recognition*, pages 354–359. IEEE Computer Society Conference, June 1992. Champaign, Illinois.

19. Chahab Nastar and Nicholas Ayache. Fast segmentation, tracking, and analysis of deformable objects. In *Proceedings of the Fourth International Conference on*

Computer Vision, Berlin, May 1993.

20. G.N Pande, G. Beer, and J.R. Williams. *Numerical Methods in Rock Mechanics*. John Wiley and Sons, 1990.

21. A. Pentland and Williams J. Good vibrations : Modal dynamics for graphics and animation. In *Computer Graphics*, 1989.

22. A. Pentland and Williams J. The perception of non-rigid motion : Inference of material properties and force. In *Proceedings of the International Joint Conference on Artificial Intelligence*, August 1989.

23. Alex Pentland and Stan Sclaroff. Closed-form solutions for physically based shape modelling and recognition. *IEEE Transactions on Pattern Analysis and Machine Intelligence*, PAMI-13(7):715–729, July 1991.

24. Lawrence H. Staib and James S. Duncan. Deformable fourier models for surface finding in 3d images. In *Proceedings of Visualization in Biomedical Computing*, Chapell Hill,USA, October 1992.

25. Demetri Terzopoulos and Manuela Vasilescu. Sampling and reconstruction with adaptive meshes. In *Proc. Computer Vision and Pattern Recognition*, pages 829–832. IEEE Computer Society Conference, June 1991. Lahaina, Maui, Hawaii.

26. Demetri Terzopoulos and Keith Waters. Physically-based facial modelling, analysis, and animation. *The Journal of Visualization and Computer Animation*, 1:73–80, 1990.

27. Demetri Terzopoulos, Andrew Witkin, and Michael Kass. Constraints on deformable models: recovering 3-D shape and nonrigid motion. *AI Journal*, 36:91–123, 1988.

28. J-P. Thirion, O. Monga, Benayoun S., Gueziec A., and Ayache N. Automatic registration of 3d images using surface curvature. In *IEEE Int. Symp. on Optical Applied Science and Engineering*, San-Diego, July 1992.

29. Manuela Vasilescu and Demetri Terzopoulos. Adaptive meshes and shells. In *Proc. Computer Vision and Pattern Recognition*, pages 829–832. IEEE Computer Society Conference, June 1992. Champaign, Illinois.

From Splines and Snakes to SNAKE SPLINES

F. Leitner, P. Cinquin *

TIMB - IMAG
Faculté de Médecine de Grenoble
38 700 La Tronche, France

Abstract. Segmenting 3-D complex medical objects from sets of parallel slices may be a difficult task. We propose a new method of active contours (or snakes) that simplifies the classical approach of snakes by embedding the intrinsic energy in the spline nature of the surface to deform. This yields a simple differential equation, that controls the evolution of the surface toward its target. The simplification thus obtained allows for taking into account automatic modification of the topology of the object to segment, as is shown on the instance of a vertebra.

1 INTRODUCTION

Active contours, or snakes, have proven a very efficient method for 2D image segmentation [1, 2], either completely automatic or semi-automatic. The generalization of these methods to 3D images puts difficult problems of complexity of the system needed to minimize the 3D generalization of the energy that defines the border of the object.

To deal with this complexity, two parallel approaches have been developed almost simulteanously. Cohen et al [3] have proposed to use finite elements to solve the 3D problem. In a previous paper [4], the authors had suggested to use a spline representation of the border of the object to segment, with a method entitled Snake Spline.

This method had initially been tested on an object with a cylindrical topology. It has since proven capable of dealing with surfaces presenting poles, such as spheres, and even to adapt to complex topology surfaces. These features require an adapted spline representation of the border of the object and an adequate strategy of evolution of the initial guess of this border to fit real data. This paper will present the characteristics of the selected representation of the surface and the solution to the adaptative deformation of the border. We will discuss in a further section the present implementation of snake splines.

2 BORDER REPRESENTATION

The border of a 3D object is a surface, which we will represent with parametric spline functions. We firstly discuss the characteristics and interest of this

* This research is partially supported by Digital Equipment Corporation

representation, then we describe its implementation, and finally we analyse the problems put by the local refinement of this representation.

2.1 Characteristics of the parametric spline representation of the border

Spline functions have for long been a popular way for representation of free form curves or surfaces [5]. These functions are interesting in the frame of active contours because most of them have variational properties [6] that make them interesting candidates for representing surfaces that are supposed to minimize some sort of "intrinsic energy".

In two dimensions, this stems very clearly out of the comparison between the energy minimized by actives contours and the energy minimized by snakes.

variational properties of spline curves Let C be the active contour, parametrically represented by

$$\forall u \in [0, 1], M(u) = \sum_{i=1}^{n} B_i(u)\alpha_i,$$

where

$$\alpha_i \in R^2.$$

The Bi are the basis spline functions. For cubic spline functions, they are piecewise degree 3 polynomials, which differ only by translations. This implies

$$\forall j \in [1, m], M_j = \sum_{i=1}^{n} B_{ij}\alpha_i,$$

where

$$\alpha_i \in R^2$$

and

$$B_{ij} = B_i((j-1)/(m-1)).$$

This can be written under a matrix form as $M = B\alpha$, where $M \in R^m$ x R^m, $B \in R^m$ x R^n, $\alpha \in R^n$ x R^n.

Finding α when M is given is a least squares problems, in which $||M - B\alpha||^2$ is minimized, thus implying $\alpha = (B^t B)^{-1} B^t M$, which we shall rewrite as $\alpha = B^{-1}M$. Let $N_i = M(u_i)$, with $u_i = (i-1)/(m-1)), i = 1, .., m$. A classical result for spline functions is that $M(u) = (x(u), y(u))$ minimizes

$$\int_0^1 [(x"(u))^2 + (y"(u))^2]du, \tag{1}$$

for all M belonging to $H^2([0, 1])$ x $H^2([0, 1])$ and satisfying $M(u_i) = N_i, i = 1, ..., n$.

The ratio n/m defines the "smoothing power" of the spline : if $n = m$, the spline is the interpolating spline. The smoothing is all the more so important as n/m decreases. This will later be an interesting feature, since it will allow for adaptative segmentation.

variational properties of snakes Active contours are not assigned to pass through given points. In fact, they are required to minimize the following energy :

$$\int_0^1 [\alpha[(x'(u))^2 + (y'(u))^2] + \beta[(x"(u))^2 + (y"(u))^2] + \gamma E(f(x(u), y(u)))]du \quad (2)$$

In this equation, f represents the initial image, and E is some sort of energy adapted to image segmentation (for instance $-||\nabla(f)||^2$), α, β and γ are parameters that can be optimized to find a compromise between bending, length and extrinsic energy.

comparison of the variational properties of splines and snakes - definition of a snake spline The expression minimized in (2) can be split into an intrinsic energy (involving the derivatives of the contour) and an extrinsic energy (which involves the image itself). As opposed to splines, the intrinsic energy involves first derivatives, which results in controlling the length of the contour. This effect is particularly interesting when, for some reason, the border of the object to segment does not attract the active contour (which may typically happen in "flat" regions of the image, where no gradient can attract the active contour). The absence of first derivative terms in (1) is not dramatic, if some technique for moving the curve in those "flat" regions is provided. We will see further that this is possible, and, moreover, in a local way. The second derivative terms are the most interesting, since they are responsible for the "smooth" aspect of the active contour. Obviously, these terms are identical in (2) and (1). This stresses the proximity between active contours and splines.

The basic idea of snake spline is therefore to minimize only the extrinsic energy, looking for the solution in an appropriate spline space which will embed minimization of some intrinsic energy. We will define a snake spline as the element of a spline space that minimizes

$$\int_0^1 E(f(x(u), y(u)))du, \quad (3)$$

where f is the image and E is an extrinsic energy. Section 3 of this paper will propose a constructive implementation of this definition.

The major advantage of snake splines over classical active contours lies in the fact that problem (3) is much simpler than problem (2) for two principal reasons :

- the dimension of the space is dramatically reduced. Indeed, the dimension of the functional space in 2 is infinity, while it can be completely controlled in 3 (the interesting adaptative consequences of this versatility will be emphasized in section 3).

– the system to solve is much simpler, since it does no longer involve the derivatives of the parametric representation of the border of the object. We will show further that it results in an Ordinary Differential Equation that can be solved very easily.

2.2 Practical evaluation of a snake spline

Each one of the numerous spline functions designed for CAD can be adapted to become a snake spline, according to definition 3. Yet the most interesting will be those that minimize an energy that has a strong "physical" meaning, or those that are computationnaly interesting. Besides, an adequate representation of the snake spline has to be selected.

Depending on the use of the spline, various representations have been proposed [7, 8]. Representations based on control vertices, initially popularized by Bezier [9] have proven very interesting from a computational point of view, because recurvsive subdivision algorithms which can easily be optimised on computers [10, 11] perfectly suit this representation. We go into more details on such a representation, briefly reviewing a 2D instance, then its generalization to 3D through tensorial products, and finally analyzing how surfaces with poles can be dealt with, because next subsection will take this representation as a basis for adding more flexibility to the spline.

2D spline subdivision algorithm Let us consider a piecewise quadratic spline, which will therefore be C^1. This spline is initially represented by a control polygon of $n+1$ vertices $\{P_i\}$, $i = 0, n$. The subdivision algorithm consists in dividing each segment $P_i P_{i+1}$ into three subsegments of equal length, thus providing $2n$ vertices $\{Q_i\}$, $i = 0, ..., 2n - 1$ (see Fig. 1). This process can be iterated. It can easily be proven that the set of points that it generates converges towards the spline. This technique can easily be generalized to higher order splines.

surface representation through tensorial products of splines When the object we have to segment lies in a 3D space, its border is a surface that we will parametrically represent by $M(u, v) = (x(u, v), y(u, v), z(u, v))$, with

$$(u, v) \in [0, 1] \text{ x } [0, 1].$$

x, y and z will be obtained by tensorial products of unidimensional splines, i.e. for instance $x(u, v) = s_u(u) \text{ x } s_v(v)$, where s_u and s_v are unidimensional splines.

The interest of these functions is that they can be very easily constructed. The previous subdivision algorithm readily applies : one just has to subdivide firstly the lines, and then the set of columns corresponding to the new mesh obtained by the first subdivision.

The drawback of these approaches is that the minimization properties satisfied by these tensorial splines usually have no physical meaning. For instance, in the case of a bi-cubic spline x, it minimizes

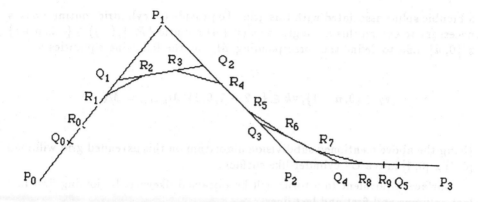

Fig. 1. subdivided control polygon $\{R_k\}$ formed from the initial control polygon $\{P_k\}$, for a second degree spline curve

$$\int_{[0,1] \times [0,1]} \left| \frac{\partial^4 x}{\partial^2 u \partial^2 v} \right|^2 \, dudv.$$

But we will see further that this regularizing energy is sufficient, and that it is not necessary to use "thin-plate splines" [12] which actually minimize the real generalization of the bending energy

$$\int_{[0,1] \times [0,1]} \left| \frac{\partial^2 x}{\partial u^2} \right|^2 + 2 \left| \frac{\partial^2 x}{\partial u \partial v} \right|^2 + \left| \frac{\partial^2 x}{\partial v^2} \right|^2 \, dudv.$$

Indeed, the corresponding spline is not separable, and is defined with Log functions. This is an important drawback for our application, since such functions have an infinite support, which makes computation time consuming, and which we will prove would be inconvenient for adaptative local refinement.

tensorial product representation of surfaces with poles The tensorial product approach is very simple to use for surfaces that can easily be mapped onto $[0, 1] \times [0, 1]$. This is of course the case of "open" surfaces with no "hole". Closed surfaces are a little more difficult to represent.

Surfaces equivalent to cylinders put no major problems. One just has to adapt the continuity contraints for the "border columns". Let us assume for instance that $[0, 1] \times [0, 1]$ is discretized by a $n \times n$ grid $\{P_{ij}\}, (i, j) \in \{0, n-1\} \times \{0, n-1\}$. Let $\{M_{ij}\}, (i, j) \in \{0, n-1\} \times \{0, n-1\}$ be a set of control vertices of a bicubic spline associated with this grid. To provide C^2 cylindric continuity, it is necessary to extend this n x n grid to a $(n+4)$ x n grid $\{P_{ij}\}, (i,j) \in \{-2, n+1\}$ x $\{0, n\}$, and to define the corresponding M_{ij} by the following equalities :

$$\forall j \in \{0, n-1\}, \forall k \in \{-2, -1, 0, 1\}, M_{n+k,j} = M_{k,j}.$$

Using the above mentioned subdivision algorithm on this extended grid will map $[0,1]$ x $[0,1]$ into a C^2 cylinder like surface.

Surfaces equivalent to a torus will be obtained likewise, by joining first and last columns and first and last lines.

Spheric topology surfaces are more difficult to represent. We will take as an example the case of tensorial second degree splines, (second degree splines were the one-dimensional instance in Fig. 1). In that case, the surface will be obtained by subdivision and refinement of a mesh of 6 control vertices (see Fig. 2). The six points P_0, P_1, P_2 and P'_0, P'_1, P'_2 can be organized as a "cylinder", using the above described technique. In our instance, this comes down to considering P_0, P_1, P_2 and P'_0, P'_1, P'_2 as two "cycles". We now have to close the second dimension. To do so, we introduce two additional control points, the North and the South poles (N and S). These control points are not real degrees of freedom, because they will be commited to be the barycenters of P_0, P_1, P_2 and P'_0, P'_1, P'_2. The tensorial structure of the surface requires that the subdivision algorithm be applied firstly in one dimension, and secondly in the other one. Let us apply the subdivision algorithm in the horizontal dimension. We obtain 2 sets of 6 points, $Q_0, Q_1, Q_2, Q_3, Q_4, Q_5$ and $Q'_0, Q'_1, Q'_2, Q'_3, Q'_4, Q'_5$. These sets have now to be subdivided along the "vertical" dimension. This requires that pairs of Q_i (resp. pairs of Q'_i) be linked through N (resp. S). To do so, "opposite" points will be linked together, thus forming $Q_0 N Q_3, Q_1 N Q_4, Q_2 N Q_5$, and $Q'_0 S Q'_3, Q'_1 S Q'_4, Q'_2 S Q'_5$. This defines vertical cycles that can be subdivided with the subdivision algorithm. Fig. 2 shows the first level of the subdivision of the cycle $\{Q_0, N, Q_3, Q'_3, S, Q'_0\}$. With second degree splines, it is obvious that this algorithm ensures C^1 geometric continuity : indeed, all the subdivision algorithm applied to each vertical cycle will always leave at least one pair of points surrounding N that belongs to plane $P_0 P_1 P_2$. This plane is therefore the tangent plane to the tensorial surface in N. It can be proven that higher order continuity can easily be obtained by the same principle, for higher order splines.

The refinement algorithm can now take place, and the surface will become more complicated, but will keep its topological equivalence with sphere. The only restriction that will be enforced is that the horizontal cycle neighbouring N should not be refined. The local characteristics of the refinement algorithm make this possible. But of course, these points will move with time : they will therefore get closer and closer to the pole, which means that the interdiction of refinement of this triangle of control points is not a limiting factor to the versatility of the snake spline in this point.

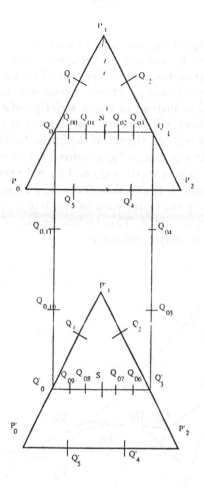

Fig. 2. first levels of the "horizontal" and "vertical" subdivision of the mesh of 6 control points used to provide spheric topology second degree tensorial snake spline

2.3 Local refinement of a snake spline

As opposed to classical splines, which are used to represent a given function, and are not supposed to evolve with time, snake splines are supposed to move progressively towards an a priori unknown shape. Besides, classical splines have a "bending energy" (corresponding to (1) that is a priori completely controlled by the number and position of the control vertices. In the case of snake splines, the desired "flexibility" is not a priori known. It can depend on local character-istics of the border (which may be "smooth" for some parts of the object, thus requiring few control vertices, or "rough" for some other parts, which would im-ply more control vertices). We therefore need a "refinement" algorithm, which will give more flexibility to the intial spline. We propose to modify slightly the initial subdivision algorithm, to turn it into the desired snake spline refinement algorithm.

In the subdivision algorithm, the control vertices are completely defined at each level of subdivision by the set of vertices of the previous level and by proportionality rules that correspond to the type of spline. To provide flexibility, the refinement algorithm will only give their "autonomy" to some selected control vertices, which means that instead of being strictly defined by proportionality rules, these control vertices will become real degrees of freedom, as the $n + 1$ initial control points. We will call the set of the initial control vertices, completed with these "freed" control points, a "generalized set of control points". Fig. 3 shows such a refinement, where vertices Q_2 and R_3 have been "freed". The rest of the algorithm is not modified. It can therefore easily be seen that the new curve keeps the same continuity characteristics as the initial spline, but that locally more flexibility has been provided. This technique can of course be generalized to spline surfaces, with no major difficulty.

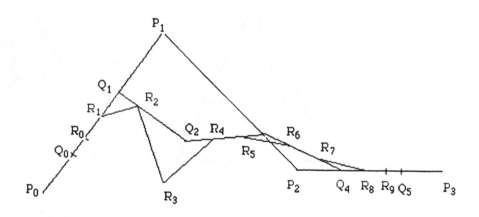

Fig. 3. refined control polygon $\{R_k\}$ formed from previous control polygons $\{P_k, Q_k, R_k\}$, for a second degree spline curve. The generalized set of control points is $\{P_0, P_1, P_2, P_3, Q_2, R_3\}$

We now have to make use of this new flexibility, which will be described in the following section.

3 EVOLUTION OF A SNAKE SPLINE IN A STRENGTH FIELD

We saw in the previous section that a snake spline minimizes (3) in the functional space defined by a generalized set of control points and by a subdivision algorithm. We shall now see how such a snake spline can be constructed, which the adaptative features of the proposed method are, and how it is possible to represent complex topology objects.

3.1 Dynamic construction of the snake spline

For the sake of simplicity, we shall present this construction in the 2D case. The tensorial product structure makes generalization to 3D images very easy. The image f in (3) is usually discrete. It is therefore not inconvenient to use a discrete version of this equation, which reads

$$\sum_{i=1}^{m} E(f(M_i)) = \sum_{i=1}^{m} V(M_i), \tag{4}$$

where $M = \{M_i, i = 1, m\}$ is the image of a set of m equidistant points in the parametric space, and where V corresponds to the potential from which the strength field derives.

To minimize 4, we would like each point M_j to move into the direction of strength $F_j = -||\nabla V(M_j)||^2$. We shall consider that the position of M_j depends upon time t. Derivating M_j with respect to t, this implies

$$\forall j \in \{1, m\}, \frac{dM_j}{dt} = F_j,$$

or, in matrix form,

$$\frac{dM}{dt} = F.$$

Let $P = \{P_i, i = 1, k\}$ be the generalized set of control points. The subdivision algorithm ensures that $M = BP$, where B is a matrix that characterizes the subdivision algorithm at the chosen precision level. P can be considered to be the solution to the minimization of $||M - BP||^2$, known to be $P = (B^t B)^{-1} B^t M$, which can be rewritten as $P = B^{-1}M$. Therefore, if M depends upon time, P also depends on t, and can be derivated with respect to t. We then obtain :

$$\frac{dP}{dt} = B^{-1}\frac{dM}{dt} = B^{-1}F. \tag{5}$$

This means that, if the curve C is put into a strength field, it will evolve with time until a balance position is found. As with the classical snake approach, two kinds of "energy" account for the final shape of C : the energy corresponding to the strength field ("extrinsic energy"), and the energy corresponding to the minimization of the sum of the squares of the second derivatives of the spline functions ("intrinsic energy"). But, unlike in the classical snake methods, it is not necessary to deal explicitly with the intrinsic energy, since it is self contained in the definition of the spline. So the resulting systems will be much easier, since they will only take into account the extrinsic energy.

3.2 Adaptative evolution of the snake spline

We have seen previously how a spline can evolve in a strength field deriving from V. Introducing V into 5, we obtain

$$\frac{dP}{dt} = B^{-1}\frac{dM}{dt} = B^{-1}F = B^{-1}V(M) = B^{-1}V(BP). \qquad (6)$$

This is an Ordinary Differential Equation, in which the unknowns are the control vertices P.

The initial values of P will depend on the *a priori* knowledge on the object. If no *a priori* knowledge is available, the initial curve will roughly fit the limits of the image. For 3D image segmentation, we will take the above mentioned spherical mesh of $n = 6$ control points as a starting point. The method will then behave adaptatively. The differential system **??** is solved with n coefficients, corresponding to n knots $M_j, j = 1, ..., n$, until they stop evolving. At that time, the refinement alogrithm is applied. Two kinds of stop criterions can be used : testing a "cost" of the contour (e.g. integral of the potential along the contour), or testing the evolution of the coefficients α. We have found the later easier to implement.

Two other adaptative features of the method, which have not yet been tested, might prove interesting. First, it is possible to adapt the number m of points on which the strengths are computed according the number of knots of the spline. Then, the kind of function g to use as a convolution with the initial potential should also depend on the number n of knots : when n is low, the curve is supposed to be far away from the correct position. So g should be broad, to increase the sensitivity of the curve to the high but potentially distant gradients. On the contrary, when n is high, g should be more narrow, to become able to fit mean gradients.

A priori knowledge can be input in the form of a CAD model of the object to segment. This has not yet been tested, but should be interesting. In that case, the first problem is to estimate as well as possible the 6 parameters (translations and rotations) for fitting the model with no modification to the real image. This is a "rigid matching" problem, for which solutions have been proposed. Once this rigid matching has been done, an "elastic matching" can take place, using the method previously described.

3.3 Complex topology object representation

The evolution of the snake spline surface can bring it to get folded in such a way that it intersects with itself. We will firstly propose a method for detecting such an intersection, and will analyze its consequences in a further subsection.

intersection detection One solution to the problem of detecting an intersection between two surfaces could consist in solving a system of equations, each line of which would represent the equation of one surface. But our surfaces are only piecewise polynomial, and this method would blow up because of its combinatory complexity. We therefore prefered to build a method of "map filling up", which is not optimal with a few patches, but has a complexity that depends only on the number of points of the object, and not on the number of its patches.

This method consists in evaluating the object point by point, and in filling up a map (tree-hierarchized and dynamically allocated for memory space reduction). Each element of this map contains the parametric coordinates of a point on its patch, and the reference of this patch. The very structure of the map codes the rounded spatial coordinates for each point of the object. So two cases may occur : if the map element is not allocated, it is filled up. If not, the map element is already filled up, and there is a possibility of intersection.

Because of the rounded coordinates (useful to increase the speed of the convergence and to decrease memory space) several neighbour points can "fall" into the same map element. That is why we have to test the parametrical coordinates of these points between themselves. If they belong to the same neighborhood (using a parametrical thresholding) we will decide there is no intersection. Moreover, because of the continuous joining between patches, the same type of problem can occur on the border of the patch. To deal with this, we build up a neighborhood graph that will be checked when a possibility of intersection appears. If the points belong to the same frontier there is again no intersection. A particular case is the problem of poles for spheres or hemispheres. In a pole, the border of the tensorial product spline representing the pole is reduced to one single point : the pole. In that case, all the points (u, v) with $u \leq L$, where L is a parametric threshold, have the same rounded spatial coordinates. We will therefore have to test wether the candidates for an intersection satisfy this property, and to discard them if they do.

If two points pass all these tests, we can say an intersection has been detected, and we know their parametrical coordinates on their surfaces.

topological transformation When an intersection occurs, it means that the initial topology of the snake spline surface does not fit the actual topology of the object to segment. It is therefore necessary to make a "hole" in the snake spline, and yet to keep its main characteristics, mainly continuity and tensorial structure.

Let S and S' be the two patches that intersect, and let O be their common point. Using the subdivision algorithm, S and S' can be subdivided until O corresponds to the center of two sets of 4 subpatches (see Fig. 4), belonging to S and S'. To make the hole, O is split into 4 points belonging to S (I, J, K, L) and 4 points belonging to S' (I', J', K', L'). These sets of points, initially located in O are progressively moved aside to "tear up" S and S'. At that stage, the two patches are perforated, but it is still necessary to connect them, in order to provide "walls" to the hole. To do so, I is moved to merge with H', J will move to B' and so on (see Fig. 5). This makes it possible to keep continuity as well as tensorial structure, as can be seen on the instance of subpatch $ABH'H$. Indeed, these features require that each vertex of the polygon be inserted in a rectangular grid. This is obvious for A, which keeps its original neigbours in S. H and B exchange their initial neighbour O with H', which belongs to S'. As for H', it is located in the intersection of a "line" and a "column" formed by groupings of half-lines and half-columns of S and S' (see Fig. 5 and Fig. 6) :

- the "line" of the new surface passing through H' consists of H' and
  - the "left" part of the line of S passing through H',
  - A' and the upper part of the column of S' passing through A'.
- the "column" of the new surface passing through H' consists of H' and
  - B and the upper part of the column of S passing through B.
  - the "bottom" part of the line of S' passing through H'.

The same process can be iterated for each of the 8 subpatches painted in grey in Fig. 5. This provides the required "walls" to the hole, as can be seen in Fig. 6, where the observer can look "inside" the hole. As will be described later, these walls will, thus enabling the surface to undergo other topological changes if required. This process of perforating the two patches, and of "darning" them, finally comes down to "reordering" the control vertices defining the surface, without modifying the subdivision algorithm used to evaluate the points of the surface.

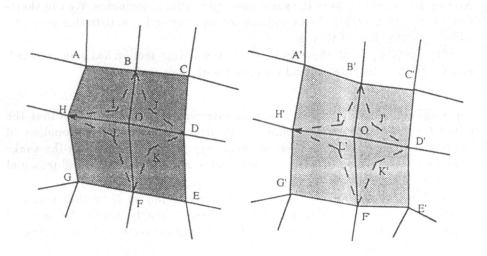

Fig. 4. The two sets of subpatches of S and S' , after adequate subdivision putting the intersection point O at the center of these sets

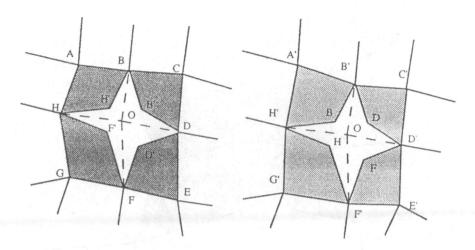

Fig. 5. Perforation of S and S' : O has been split into 8 points that have been moved towards adequate "targets" in S and S' to "darn" these two patches

4 PRESENT IMPLEMENTATION OF SNAKE SPLINES

The border representation and its evolution in a strength field presented in the previous sections have been implemented and tested on various sets of medical images, which we will present. This has lead us to use specific strength fields, rather than the classical gradients. The results will be described.

4.1 Characteristics of the 3D medical images to segment

We have tested our method on sets of parallel slices of Computed Tomography (CT) or Magnetic Resonance Imaging (MRI). These images presently provide typical $2D1/2$ images, in the sense that the resolution is much better in the two directions corresponding to the slices than in the third dimension, which is normal to the planes of the slices. The CT slices were $1mm$ thick and discretized in 512 x 512 pixels, each of which about $0.25mm^2$ in surface. we processed 50 such slices.

We were interested in surfaces that can easily be seen on these images : the border of the vertebra.

4.2 Gradient based construction of the strength field

We firstly computed a Sobel gradient on each CT slice, and then convolved it with functions such as

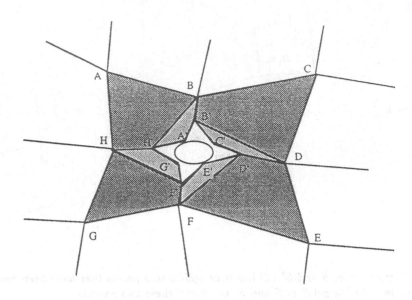

Fig. 6. Actual position of the eight subpatches that form the walls of the hole : inside view of the hole

$$g(r) = \begin{cases} 1 \text{ if } r < 1 \\ \frac{1}{r} \text{ otherwise} \end{cases}$$

In fact, this approach leads to ambiguities when two contours are close, because the active contour is attracted both contours. This drawback corresponds to the fact that the least squares first approximations of the contours are not very flexible (since they are defined by a limited set of control points) and may therefore cross the border and later get attracted by other structures.

These considerations lead us to use another approach.

4.3 Distance map based construction of the strength field

The strength field that is used to attract the snake spline surface can be simplified by a pre-processing that will provide indications about points that are candidate to belong to the border of the object. This pre-processing step, which uses classical tools, transforms the initial volume of data into distances from each point of the snake spline to the nearest boder point. The chain of processing begins with local operators (edge detectors) and proceeds with a regional analysis.

edge detection It is performed with classical edge detectors, such as Sobel gradient thresholding or zero crossing of Laplacians. Of course, these 2D operators are not optimal for 3D segmentation, but this is not dramatic in our case, since we do not need a perfect identification. Indeed, we just need indications about points which belong to the border with a high degree of probability. The rigidity of the snake spline will propose a continuous guess between these points.

Besides, multi-scale computation of these edges can be used : during the first steps of snake spline deformation, the edges are computed after convolution of a large scale gaussian, and this scale is progressively reduced when the snake spline converges towards its aim.

identification of border elements To reduce noise and to make computation of the distance between the snake spline and the borders easier, we proceed to a regional analysis in four steps.

contour thinning The first step consists in getting a contour of one pixel of thickness, which is necessary when the edge points have been obtained by gradient thresholding. This is performed by selecting only crest points (i.e. points whose gradient is maximal in a direction perpendicular to that of the gradient).

proximity map building The second step "expands" the edges voxel by voxel, with a distance flag. This technique is computationaly very efficient, and uses dynamic programming [13]. It does not provide a euclidian distance, but this will not bother us, because we will see that this distance is used in a quasi qualitative way. We therefore call the result of this processing a proximity map.

border building The third step aims at linking edge points. We will show that it is not necessary to perform this linking on all edge points, and then briefly describe how they are connected.

To save computing time and to avoid processing parts of the image which will not "attract" the snake spline (because they are too distant from it), this step is initiated by the points of the snake spline themselves. Indeed, for each point P on the snake spline, we need to know its nearest border neigbour. The proximity map will be climbed down from P (simply by choosing as the following point on the path the one with minimal proximity), until a border point is reached. When two candidates have the same proximity, one of them is arbitrarily choosen, which will not affect the final result (because it is likely that the neighbour of P on the snake spline will not go through the same ambiguity, and the deformation of the snake spline will therefore be globally coherent).

Once a starting edge point has been selected, a classical contour tracking is performed on its neighbour edge points. A cooperation between gradient thresholding and zero crossing of the laplacian is performed : holes can occur between edge points defined by gradient threseholding. These holes will be "filled" by tracking the zero crossings of a Laplacian, which of course have no hole. The advantage of this approach is that the comnputation of the zero crossings of the Laplacian, which is computer power consuming, is performed only locally, when it is really useful.

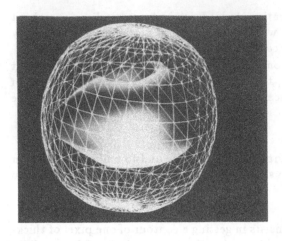

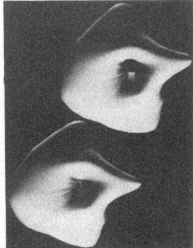

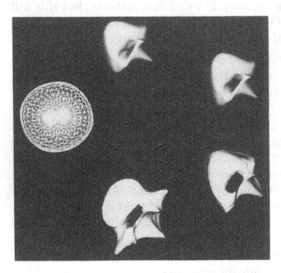

Figure 7 : Initial surface and first deformation (top). Two intermediate deformations of the surface, showing the appearance of the hole (middle). Clockwise, 6 steps of the evolution (bottom).

border verification The advantage of this regional level of processing is that it is possible to reduce noise by keeping only those borders that meet specific criteria. The simplest criteria are geometrical ones : we will keep only those borders that have a size greater than a given threshold (higher level criteria, such as integral of the sqare of the second derivatives, could also be considered).

But the position of the snake spline and the local characteristics of the image may also be taken into account. Indeed, each border element can be associated with a parameter that describes the border and its neighbourhood. We are presently using the mean value of grey levels along each edge. This value is supposed to be consistent between one border element and its neighbour border elements. The variance of this parameter, computed on one border element and its immediate neigbours, should therefore not exceed a given threshold. This of course implies to be able to define an order on the border elements. This order is given by the snake splines : two border elements are neighbour if they are connected to neigbour parts of the snake spline (this connection is defined by the path originating from the snake spline and that has reached one point of the border element).

These two techniques provide elements to decide whether border elements should be discarded. When such a decision is taken, the proximity map must of course be accordingly modified.

4.4 surface evolution

The availability of the distance map provides an alternative to the resolution of the differential equation presented in section 3. Indeed, for eah point of the ative contour, we can easily know, thanks to the distance map, which the nearest edge is. We now have to move the control grid to reach this edge. The hole management and surface refinements are dealt with as described in section 3.

Fig 7 shows the result on an image of a vertebra : the initial surface, globally spherical, is progressively transformed into the actual vertebra.

5 CONCLUSION

This *SnakeSplines* method has proven very efficient for complex topology object representation and segmentation. It presents interesting features in terms of hierarchical representation of an object, that make it an interesting candidate for interactive design of complex surfaces. Besides, it has intrinsic properties of parallelism which allow for massively parallel implementation. We are presently working on these two directions.

References

1. Demetri Terzopoulos, Andrew Witkin, and Michael Kass. Constraints on deformable models: recovering 3d shape and nonrigid motion. *AI Journal*, 36:91–123, 1988.

2. Demetri Terzopoulos. On matching deformable models to images. In *Topical meeting on machine vision, Technical Digest Series*, volume 12, pages 160–163. Optical Society of America, 1987.

3. Laurent D. Cohen and Isaac Cohen. A finite element method applied to new active contour models and 3d reconstruction from cross sections. In *Proc. Third International Conference on Computer Vision*, Osaka,Japan, December 1990.

4. S. Leitner, I. Marque, S. Lavallee, and P. Cinquin. Dynamic segmentation : finding the edge with spline snakes. In P.J. Laurent, A. Le Mehaute, and L.L. Schumaker, editors, *Curves and Surfaces*, pages 279–284. Academic Press, Boston, 1991.

5. A. Le Méhaute P. J. Laurent and L. L. Schumaker. *Curves and Surfaces*. ACADEMIC PRESS, 1991.

6. P. J. Laurent. *Approximation et Optimisation*. HERMANN, Paris, 1987.

7. C. de Boor. *A practical guide to splines*. SPRINGER VERLAG, Berlin, 1978.

8. J. C. Beaty R. H. Bartels and B.A. Barsky. *Mathématiques et CAO, B-Splines*. HERMES, 1987.

9. P. Bézier. *Mathématiques et CAO, Courbes et Surfaces*. HERMES, 1987.

10. E. Cohen, T. Lyche, and R. F. Riesenfeld. Discrete b-splines and subdivision techniques in computer aided design and computer graphics. *Computer Graphics and Image Processing*, 14:87–111, 1978.

11. P. Sablonnière. Spline and bézier polygons associated with a polynomial curve. *Computer Aided Design*, 10:257–261, 1978.

12. J. Duchon. Interpolation des fonctions de deux variables suivant le principe de la flexion des plaques minces. *RAIRO Analyse Numérique*, 10:5–12, 1976.

13. Y. Robert S. Miguet. Path planing on a ring of processors. *Int. J. of Computer Maths.*, 32:61–74, 1990.

Printing: Weihert-Druck GmbH, Darmstadt
Binding: Buchbinderei Schäffer, Grünstadt

Lecture Notes in Computer Science

For information about Vols. 1–629
please contact your bookseller or Springer-Verlag